Fundamentals of

Agribusiness
Finance

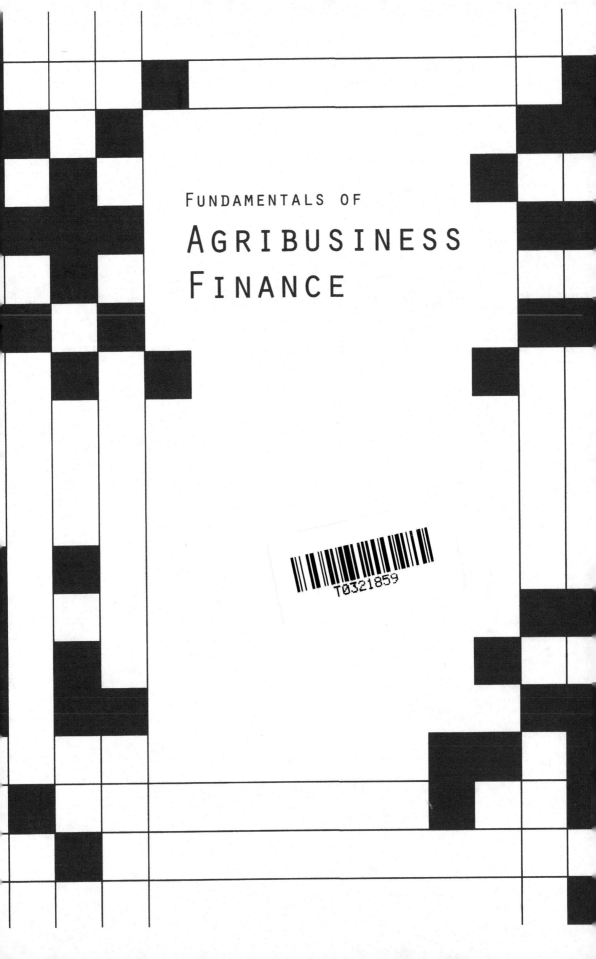

Fundamentals of
Agribusiness
Finance

Ralph W.
Battles
and
Robert C.
Thompson,
Jr.

Iowa State University Press / Ames

Ralph W. Battles received his MBA in finance from California Polytechnic State University, San Luis Obispo. Professor Battles is a university lecturer in corporate and agricultural finance at California Polytechnic State University and was Finance Faculty of the Year, 1996, 1997, and 1998 in the College of Business. In addition, Professor Battles's research interests include primary and secondary mortgage markets, agricultural lending and secondary markets, and hedging techniques for debt securities. In addition to his academic responsibilities, Battles is a financial planner in the business community.

Robert C. Thompson, Jr., received his PhD in agricultural economics at Colorado State University, Ft. Collins. In addition to his teaching responsibilities at California Polytechnic State University, Dr. Thompson has taught agricultural economics, agribusiness finance, and agribusiness managerial accounting in Colorado, Mexico, and England.

Iowa State University Press
2121 South State Avenue, Ames, Iowa 50014

Orders: 1-800-862-6657
Office: 1-515-292-0140
Fax: 1-515-292-3348
Web site: www.isupress.edu

Authorization to photocopy items for internal or personal use, or the internal or personal use of specific clients, is granted by Iowa State University Press, provided that the base fee of $.10 per copy is paid directly to the Copyright Clearance Center, 222 Rosewood Drive, Danvers, MA 01923. For those organizations that have been granted a photocopy license by CCC, a separate system of payments has been arranged. The fee code for users of the Transactional Reporting Service is 0-8138-2069-3/2000 $.10.

⊗ Printed on acid-free paper in the United States of America

First edition, 2000

Library of Congress Cataloging-in-Publication Data

Battles, Ralph W.
  Fundamentals of agribusiness finance/Ralph W. Battles and
  Robert C. Thompson, Jr.—
  1st ed.
      p.   cm.
  Includes bibliographical references and index.
  ISBN 0-8138-2069-3
  1. Agriculture—Finance. 2. Agricultural industries—Finance. 3. Agriculture—Economic aspects. 4. Agricultural credit. 5. Farm mortgages. I. Thompson, Robert C. (Robert Clarence) II. Title.
HD1437 .B38 2000
630′.68—dc21   00-038326

The last digit is the print number: 9 8 7 6 5 4 3 2 1

Contents

Preface

Fundamentals of Agribusiness Finance grew out of the authors' and others' need for an appropriate undergraduate finance text for students entering the agribusiness profession. The text is designed to cover all basic topics, including time value of money, agricultural lending, financial statement analysis consistent with Generally Accepted Accounting Principles (GAAP), the Farm Credit System, risks in agribusiness, legal issues, and nationwide and global trends in agribusiness finance.

Agricultural firms differ from most major corporations in several areas. Most agribusiness is privately owned; major corporations are owned by shareholders and shares are usually publicly traded. Although farm sizes have been increasing for several decades, the U.S. landscape is sprinkled with smaller family farms, owned by multiple generations of hands-on growers, with little formal finance training and minimal access to financial resources. This text is designed to prepare both the new graduate returning to the family business and the agricultural corporation new hire to begin making sound financial decisions by putting the fundamental principles into practice.

Within the state university or college agribusiness curriculum, this text is viewed as entry-level reading. Chapters are set forth in the order the authors have found appropriate for the entry-level course. Prerequisites are minimal; some financial statement familiarity and basic math/calculator skills should be sufficient. Where appropriate, the authors have provided sample and end-of-chapter problems for student practice and homework. The text is designed to be covered in a ten- or eleven-week quarter. For semester use, the instructor can devote more emphasis to "tougher" topics such as capital budgeting and financial-statement analysis.

The authors are indebted to their colleagues for their assistance and support, particularly those who have taught a similar entry-level agribusiness finance course, the team at Iowa State University Press, Tim Gulliver of Central Coast Farm Credit, Sue Olson for her invaluable technical assistance, and Edith Battles for her editorial prowess and periodic morale boosts. A successful textbook is truly a collaborative effort. The authors hope you find *Fundamentals of Agribusiness Finance* most useful and the starting point for your further studies in finance.

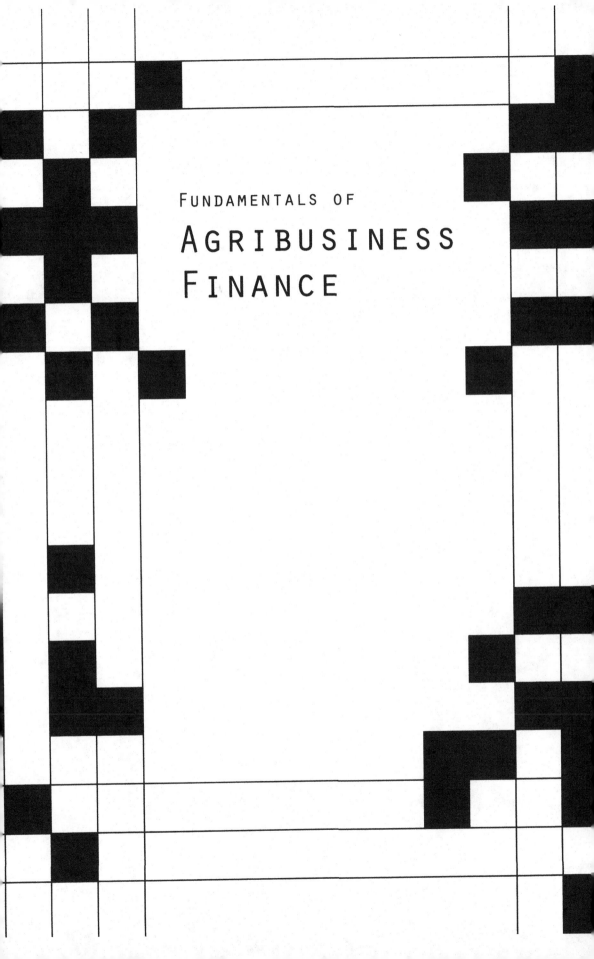

FUNDAMENTALS OF

AGRIBUSINESS
FINANCE

1

Introduction to Agribusiness Finance

Agribusiness Finance
Agribusiness Producers
Flow of Funds in the Economy
Importance of Finance to Agribusiness
Domestic Trends in Agribusiness Finance
International Trends
Questions and Problems

In the twenty-first century, agriculture is an evolving field, in both technology and business practices. As producers take advantage of advancements in biotechnology and genetics, so should they utilize today's business practices to help ensure their success. One key to this success is understanding agribusiness finance.

■ Agribusiness Finance

Agribusiness encompasses the activities of supplying goods and services to growers and ranchers, as producers providing food and fiber products, and the marketing of these products to the end customers, households and businesses. The end market for the agribusiness is the consumer. This broad definition and focus should serve us well in studying agribusiness finance.

Finance as a noun is the science of monetary affairs, and as a verb usually means to supply money for.[1] Finance professionals tend to think of it as managing money. Perhaps the most useful definition for academic purposes is the study of the flow of funds in an economy or firm.

For our use throughout this text, a convenient working definition is managing the flow of funds in an agribusiness. The purpose of business, agribusiness included, is to make money, so by substituting "money" for "flow of funds" above, we have a better, reasonably concise working definition: managing the money in an agribusiness. Of course, this does not limit our interest to the single firm; the agribusiness operates in the microeconomy but must utilize financial resources available from the macroeconomy, where it obtains the money or funds it needs to operate. These resources can be viewed as sources of funds and their utilization by the agribusiness as uses of funds.

■ Agribusiness Producers

One way to view agribusiness is to segment the producers into six categories: (1) lifestyle farmers (who don't need the income), (2) subsistence farmers, (3) growing midsize farmers, (4) midsize farmers who are scaling back operations, (5) industrial family farms generating one million dollars to twenty million dollars in sales, and (6) megafarms such as major corporations. These businesses, along with the suppliers of goods and services and the marketers of the products, are the key users of funds.

As a result of increased urbanization, some producers are operating in increasingly regulated environments. This strains the production operation for two reasons, legal constraints and potential liability. At the same time, new technology is making the role of the grower more sophisticated, perhaps increasing the producer's awareness of the business environment, including financial alternatives. Certainly, the modern producer should be trained in agribusiness finance.

■ Flow of Funds in the Economy

Although capital markets have been evolving technologically and securities such as derivatives have become increasingly sophisticated, the key players in the United States' macro, or overall, economy remain in one of three categories. These three categories are the savers or investors of capital, the borrowers or users of capital, and the financial markets and institutions that facilitate the flow of funds from investors to users. Figure 1.1 indicates that a key source of capital or funds for the macroeconomy is individual households. Families and individuals who set aside some portion of their disposable income in the form of savings provide the major source of funds for the users of capital, including agribusiness users. Other sources of funds are business savings and government surpluses when they occur. Agribusiness users access these funds through the financial markets; specifically, financial intermediaries and "middlemen," a non-gender-specific class of market makers that derives its revenues from sales commissions. Financial intermediaries such as commercial banks accept savers' deposits and invest them in loans and other securities, profiting on the interest rate differential. Middlemen merely derive a commission with each transaction, such as the sale of stock or bonds.

Figure 1.1 Source of capital for the macroeconomy.

How does the agribusiness macroeconomy differ from that of the non-agricultural corporate sector? Are the investors the same households, individuals, or businesses that provide funds for public corporations? How do the financial intermediaries and middlemen differ? Finally, how big an impact does financing agribusiness have on the macroeconomy?

The typical household does not make direct investment through a middleman. Even the rapid growth in the mutual funds industry has been accomplished through the workings of an intermediary—the mutual fund itself. The principal financial contact of most households and individuals is with a depository institution, that is, a bank, savings and loan, or credit union. These intermediaries provide a majority of the capital for the users of funds, including agribusiness. Commercial banks such as Wells Fargo and Bank of America are key lenders to farms and other agricultural businesses.[2] Banks, the Farm Credit System institutions, and life insurance companies provide mortgage loans to family farms. The Federal Agricultural Mortgage Corporation, better known as **Farmer Mac**, then purchases a portion of these mortgages and provides a method of "securitization" that allows other investors such as individuals and pension funds the means to purchase a share in a pool, or collection, of these loans.

So the agribusiness or "ag" sector of the macroeconomy operates much like the larger corporate sector, with a major role played by depository intermediaries, and a developing "secondary market" that consists of the sale and securitization of ag loans with the help of Wall Street and Farmer Mac. The role of households and individuals, having placed their savings at their commercial bank, is no different than for those depositors whose bank provides funds for nonagribusiness borrowers, nor do the savers concern themselves with the profitability of the bank's investment. But with total agricultural loans outstanding of less than two hundred fifty billion dollars,[3] the ag sector is a relatively minor part of the macroeconomy.

■ Importance of Finance to Agribusiness

The total value of all new loans in the United States for 1997 was $2.9 trillion with agribusiness comprising $43.9 billion.[4] The fact that loans to agribusiness borrowers make up less than 1.5% of the total borrowing in recent years doesn't diminish the importance of finance to agribusinesses themselves.

However, debt finance for agribusiness is not as important as **equity capital**, the owner's investment in the business. This reflects the rather conservative nature of food/fiber producers and agribusiness lenders. Compared to corporate borrowers, agribusinesses hold a larger proportion of their collateral in land, as opposed to shorter-term assets such as manufacturing equipment. In recent years, agricultural real-estate loans have dominated total agribusiness loans, at 54% of the total loans outstanding.[5] In this text, we address the use of equity capital through our discussion of capital budgeting in chapter 4, while loans and lending comprise chapters 2 and 3. In short, knowledge of both debt and equity financing is crucial to the ability of students preparing for an agribusiness career, and practitioners who financially manage agribusiness firms.

■ Domestic Trends in Agribusiness Finance

Although the 1980s saw trends of reduced farm income and reduction in farm credit outstanding, the 1990s have seen a reversal of both. While not overly burdening the industry, farm debt is on the rise and record income was reached in 1996.[6] Agricultural property consolidation continues, with private and public corporations gaining market share. **Farm cooperatives**, the joint ownership of nonprofit marketing or processing systems by producers, though struggling somewhat, continue to make headway nationally with 34% of the market for food-fiber products sold in 1996.[7] In California, increased acreage has been devoted to crops such as wine grapes and specialty fruit as growers attempt to meet changing consumer tastes. Fruit and vegetables are gaining along with dairy production, while field crops are in decline. In spite of climatological disruptions such as the nineties' El Niño and drought, national crop production continues trending upward.

The key market for agribusiness loans has been the expansion-minded midsize farmer. However, with farm consolidations and corporate farm growth, this key market is declining. To finance this market, the financial institutions have been competing for borrowers, with the **Farm Credit System**, a nationwide network of legislatively founded agribusiness lending agencies, continuing to gain market share. With the reorganization and capitalization of Farmer Mac, the financial market for loan-backed securities continues to evolve providing a secondary market, or resale market, for new agribusiness loans, similar to our national home mortgage secondary market. By learning from the experiences of the three home mortgage Government Sponsored Enterprises known as Fannie Mae, Ginnie Mae, and Freddie Mac, Farmer Mac should continue to gain sophistication and market share of new ag loan sales.

■ International Trends

With the spread of capitalism throughout the world, the recent Asian recession, trouble spots in Latin America and the former Soviet empire, the interconnectedness of world stock and credit markets, and the globalization of communication and trading, we should expect to see significant changes in agribusiness finance.[8] Most likely, younger farmers will adapt to and adopt the computer-based technology available and will be able to search for financial alternatives over the internet, much as a home buyer can search for a mortgage today. Purchasing power in the form of cooperatives and trade associations should facilitate cost savings and additional sources of financing. Foreign banks continue to expand within the United States, while domestic banks consolidate. In the future, the producer will be able to access funds either through the local community bank, larger international banking sources, the Farm Credit System, or perhaps a national trade association.

As government price controls and subsidies decline in importance, derivative securities will provide the means of controlling price risk. The Chicago Board of Trade listed the first commodities futures in the mid–nineteenth century, while futures markets are spreading worldwide to manage price risk. Producers need a rudimentary understanding of specific derivatives trading to manage this risk.

It is a modern world for the agribusiness, and the need and ability to adapt to modern financial methods and technology are key to the profitable survival of the operation. In future chapters, we study the methods and techniques to attain that successful agribusiness financial operation.

■ Questions and Problems

1-1. Consider the following career finance positions and determine if each position exists as an intermediary, a middleman, or both.

   a. stockbroker
   b. loan officer
   c. checking and savings account sales representative
   d. life insurance salesman
   e. savings and loan president

1-2. In 1997, loans to agribusiness made up what percentage of the total new loans in the United States?

■ Notes

1. Funk & Wagnalls, *Standard College Dictionary* (California State Department of Education, 1967).
2. *Journal of Agricultural Lending* 11, no. 3 (spring 1998).
3. U.S. Department of Agriculture, Economic Research Service, *Agricultural Income and Finance Situation and Outlook Report,* February 1997.
4. Federal Reserve Bank of San Francisco, December 31, 1997.
5. U.S. Department of Agriculture, *Economic Indicators of the Farm Sector— National Finance Summary,* 1991.
6. U.S. Department of Agriculture, *Economic Indicators.*
7. Charles Kraenzle, Telephone interview, July 27, 1998.
8. Daniel Klinefelter, "The Future of Ag Lending," *AgriFinance,* February 1998.

2

Agribusiness Loans: Legal Issues, Terms, and Interest Rates

Modern corporations utilize sources of capital that include the contributions of owners, the retaining of prior years' earnings, and borrowed funds from a variety of sources. Agribusinesses likewise can access both owners' contributions and borrowed funds. This chapter discusses the legal, contractual, and economic issues of borrowed funds.

■ Promissory Note

Although owners' equity dominates the capital for agribusiness, borrowed funds through the credit markets provide the liquidity that growers, marketers, and suppliers need for routine operations. The contract that stipulates the conditions of the loan is known as the promissory note.

A **promissory note** is written evidence of a debt. In essence, it is the loan document that creates a contract between the parties. As a contract, it must have several elements to be effective. These elements include the names of the borrower and lender, a promise to pay a specific amount, the due date of the note and manner of payment, the interest rate, the naming of any collateral, penalties for late or nonpayment, whether it can be prepaid, the date of the initial creation of the note, and the signature of the borrower.[1] Because of these elements, promissory notes are available in "boiler plate" form, such as from a stationary supplier or escrow company. Depository institutions have their own formats that include these elements and others specific to their policies. Since notes are negotiable instruments, like personal checks, they can be sold. In fact, there is a huge market in the United States for the resale of note instruments, known as a **secondary market**.

■ Parties to Loan Transactions

The parties or participants in an agribusiness loan transaction include the borrower, the lender, and, sometimes, third parties who assist in the transaction. The borrower is the individual, group, or business that becomes the debtor and must make repayment of the debt under the terms of the promissory note. The borrower is known as the **maker of the note.** The lender provides the initial funds to the borrower and can be a private party, a commercial bank or other depository institution, a government-sponsored agency, or other business. The lender is known as the **beneficiary** of the note, as it receives the benefit of the principal and interest payments. Third parties include escrow companies that assist in the initial transaction, dealers or vendors who provide personal property to the borrower, and possible asset-recovery entities, such as the local court system, when the terms of the note are not satisfied by the borrower. For personal-property financing such as equipment, third-party enforcement of the note is not normally required. The lender can take action to recover the collateral when the borrower is in violation of one or more terms of the note.

■ Security Agreement

Under the Uniform Commercial Code, national guidelines adopted by states for the legal conduct of business, lenders are able to specify the collateral for a loan through the use of the **security agreement**. This document is a standardized description of the assets pledged as collateral for the loan. The security agreement should contain elements similar to the promissory note, such as identification of the parties and the indebtedness, and also a precise description of the collateral or "security" for the loan. The collateral is classified by type of asset, such as inventory or equipment, and the agreement is signed by both the borrower and the lender. Sometimes the agreement is augmented with an additional document such as the warehouse trust receipt, specifying certain inventory on the premises of the borrower that cannot be sold without consent of the lender.

To properly utilize the security agreement, the lender should file a **financing statement**, a short summary of the transaction and collateral, at the local county recorder's office for real property and with the secretary of state for personal property. This provides a public record of the lender's interest in the collateral and enables recovery of the collateral if the borrower defaults on the promissory note.

■ Mortgages Versus Deeds of Trust

For real property transactions, collateral is provided by the real estate. In the United States, two systems of security agreements prevail to demonstrate an interest in the collateral by the lender. Most of the country subscribes to the **mortgage** document, a pledge of the real estate to the lender as security for the debt. In California and several other western states, the **deed of trust** is preferred. It provides temporary title to the collateral to an independent third party, usually a title or escrow company. This weak form of ownership interest in the collateral only has effect if the borrower fails to comply with the promissory note, that is, goes into **default**, the failure of the borrower to meet the terms of the loan contract.

Parties to real-estate loan transactions differ according to whether the loan is created in a mortgage or deed-of-trust state. Table 2.1 differentiates the parties under the two systems.

Table 2.1. Parties to real estate loans

Party	Trust deed states	Mortgage states
Borrower	Trustor	Mortgagor
Lender	Beneficiary	Mortgagee
Third party enforcement	Trustee	Civil court

■ Other Loan Terminology

Most commercial loans to businesses and consumers require repayment of **principal**, the amount originally borrowed or owed to the lender, and **interest**, the "rent" charged by the lender for use of the principal. The monthly or other periodic payment can be interest only or interest and principal, and may be insufficient to pay the current interest earned by the lender, therefore causing the principal balance to increase over time.

The **loan term** is the length of time until the loan contract matures. Short-term loans have terms of one year or less. In agricultural lending, loans with terms of between one and ten years are often called intermediate-term loans.

Long-term loans are those that mature in more than ten years. The phrase "term debt" refers to the combination of intermediate-term and long-term debt, that is, any loan with a term of more than one year.

Amortization is the periodic payment of the principal to extinguish the debt over the loan's term. Most agribusiness loans are amortizing or **fully amortizing**, meaning the periodic payments are sufficient to extinguish the debt over the term of the loan. A **partially amortizing** loan is one in which the loan payments make some reduction of the principal, but are not large enough to extinguish the debt over its term. Partially amortizing loans ultimately have a **balloon**, or lump sum, payment of the principal at the end of the loan's term. More common than partially amortizing loans are interest only loans, in which the payment only covers the current interest earned by the lender on the loan principal.

The **annual percentage rate**, or APR, is the actual or effective rate of interest for the loan, given all the requirements including the stated interest rate, any prepaid interest or other lender fees, and the timing of all the payments. The effect of these fees is to raise the interest rate to the lender. In 1968, Congress passed the first Truth in Lending Act, directing the Federal Reserve Bank System to create regulations that would require federally related lenders to disclose the APR. By 1970, the Federal Reserve had issued Regulation Z, which was designed to protect consumers from lenders who would quote one loan interest rate and then actually earn a higher one.

The prepaid interest that many loans require is known as **points**. One point is 1% of the loan principal. If you borrow $500,000 to finance your vineyard, and the lender charges two points, this will be $10,000 in prepaid interest. Points do not reduce the interest paid over the term of the loan; they merely enhance the yield earned by the lender, and were one justification for the Fed's Regulation Z. We will determine the effect of points on a loan's yield in chapter 3.

An amortization schedule is an algorithmic table that details the repayment of the loan principal and the payment of interest for the entire term of the loan. The amortization schedule for a $100,000 loan with a five-year term, an interest rate of 10%, and annual payments is shown in table 2.2. Note how the principal balance drops to zero at the maturity of the loan.

Table 2.2. Amortization schedule

Year	Beginning balance	Payment	Interest paid	Principal paid	Ending balance
1	$100,000	$26,360	$10,000	$16,380	$83,620
2	83,620	26,380	8,362	18,018	65,602
3	65,602	26,380	6,560	19,819	45,783
4	45,783	26,380	4,578	21,801	23,982
5	23,982	26,380	2,398	23,982	0

The term of the loan usually dictates whether it requires interest only or amortizing payments. Most long-term and intermediate-term loans are fully amortizing, while short-term loans tend to be interest-only or single-payment loans at maturity. Long-term loans tend to be for the purchase of real estate, while intermediate-term loans are usually for the purchase of equipment or other personal property, such as breeding livestock.

For agricultural real estate and private home ownership, loans are categorized by size. Private home loans greater than $227,150 are labeled jumbo mortgages, while loans below that figure are considered conforming. Annually, the Federal Housing Administration, in conjunction with the Federal National Mortgage Association (FNMA) and the Federal Home Loan Mortgage Corporation (FHLMC), sets the limit for conforming home loans. These federally sponsored agencies consider conforming mortgages to be typical middle-class borrowings with a high probability of repayment.

■ Life Cycle of a Loan

The seven phases to the loan process are shown in figure 2.1.

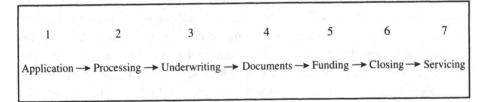

Figure 2.1   Life cycle of a loan

1. The initial step to any loan is the *application* phase, or the "building" of the loan file. For federally sponsored agencies, loan application forms are standardized, such as the FNMA/FHLMC four-page home-loan application. Most commercial lenders utilize a similar form. The data gathered on the application are used to enable the **underwriter**, the individual, software program, or department making a determination as to whether or not to grant the loan, to have sufficient information to make that determination, particularly when it is combined with the borrower's credit history. Other application-phase documentation includes a real- or personal-property appraisal, the borrower's formal credit report, and numerous verifications such as employment, tax returns, insurance, and other property holdings and indebtedness.

2. Once the paperwork or computer file is gathered, the loan enters the phase of *processing*. Still early in the cycle, the processor, frequently the loan agent, gathers the materials requested in the application phase. A formal

hard file or computer file is prepared, any remaining verifications are requested, and the loan file is organized to facilitate the next phase, underwriting.

3.   The third phase of the loan process is crucial to both the lender and the borrower. In *underwriting*, the loan file is scrutinized so that a determination can be made to either approve, deny, or suspend the request. Criteria to make this determination include the applicant's credit history and explanations for any late or unpaid payments, credit scoring such as the FICO or Fair-Isaacs numerical rating, review of employment or other income verifications, and key-ratio analysis. A variety of financial ratios is utilized. We will address these ratios in chapter 6.

Because underwriting can be more art than science, the underwriter may approve a file that does not meet one or more of the lender's ratio criteria. In working for the lender, underwriters attempt to approve loans whenever possible, and may use **compensating factors**, financial and personal issues that positively affect the likelihood that the borrower will meet the obligation, to justify a favorable ruling on the application. For example, borrowers who have slightly deficient income may gain approval due to a higher initial investment or because they are nearing the expiration date of a major monthly payment such as automobile financing. But the computerization of lending is gaining acceptance in the industry, and the day may soon come when routine loan approval is provided without the subjective human element. For example, FNMA has an on-line automated underwriting system, its Desktop Underwriter, which eliminates the need for a paper file to be submitted, and replaces the manual underwriter's position.[2] Larger loan amounts may require approval of a loan committee or asset/liability committee. Commercial banks usually give their loan officers discretion to underwrite and approve loans below specific dollar limits, such as five hundred thousand dollars. For loans above this limit, the officer must present the file to the loan committee or to a supervising loan officer.

Some applications are considered too weak to approve with the existing information. If the underwriter believes the file still has merit, he or she may suspend the approval, that is, withhold a final determination of approval or denial until further evidence is presented. Suspense is common when there is a lack of sufficient evidence of income, a sketchy credit history, collateral valuation difficulties, unexplained indebtedness, or other missing information. Outright denial of a loan request usually is not reversible for the borrower or loan officer until the borrower's or lender's circumstances have changed. Poor credit history and lack of sufficient income are frequent causes for denial. Assuming the loan is approved in underwriting, the next phase is the preparation of documents.

4. Preparation of contractual *documents*, or "docs," usually immediately follows underwriting approval. A formal security agreement for the loan collateral, or mortgage or deed of trust for a real-estate loan, must be prepared, along with the promissory note. Truth-in-lending documents, such as the APR disclosure, and antidiscrimination information, such as ECOA disclosures from the Equal Credit Opportunity Act, are routinely prepared for real-estate loans. Ultimately, the borrower may review and sign as many as ten pages of documents for a short-term loan and thirty or more pages for a real-estate loan.

   Frequently, document signing by the borrower takes place at a third-party location such as an escrow company. Real-estate documents such as deeds require the signature of a notary public, usually provided by the escrow company. Non-real-estate loans may not require a notary and can be executed at the office of the commercial lender.

   For some loans, document preparation is put on hold following the underwriter's loan approval until some future event, such as the sale of an asset by the borrower, the progress of a coincident escrow, or the fixing of the loan-interest rate, which was undetermined at the time of the application. When a borrower applies for a loan but the loan interest rate is not contractually guaranteed, the rate is said to be floating. Of course, variable-interest-rate loans are not fixed during the term of the loan. If, however, the lender has guaranteed the borrower a specific interest rate prior to the loan approval, then the rate is said to be locked. Since the promissory note is one of the key documents to be executed at signing, and since it must indicate the interest rate for a fixed-rate loan, it cannot be prepared until the interest rate is contractually guaranteed by the lender. This puts a "hold" on document preparation and execution.

5. When loan documents have been properly prepared and executed, they are returned to the document preparation unit of the lender by the escrow company or loan agent and reviewed for completeness and legal compliance. If all is in order, then the lender is prepared to fund the loan, placing a deposit into escrow or preparing a cashier's check for the borrower. If the *funding* goes to an escrow, the escrow company will guarantee that the funds are not released to the borrower until all contractual issues have been resolved, such as recording a mortgage or deed of trust in favor of the lender or trustee, or providing evidence of the borrower's insurance to the lender. If the borrower is to receive funds directly from the lender, the funds will not be released until the borrower has met all contractual obligations. Frequently, funding goes to a third party and never actually is received by the borrower. In a real-estate purchase, the seller and other parties would receive the funds. For a personal-property purchase such as farm equipment or a new vehicle, the dealer/seller will be the recipient of the proceeds.

6. The formal release of funds to the appropriate party such as the seller of the property or equipment dealer occurs at closing. For a real-estate transaction, this will immediately follow the recording of the appropriate deeds and mortgage or deed of trust. For vehicles, it occurs when legal title is provided to the lender, while the registered owner of the vehicle will be the borrower.

7. The final phase in the life cycle of a loan, **loan servicing**, the maintenance of payment, tax, and other records for the lender and borrower, is frequently performed by the original lender. The servicer also makes certain that any payments are passed through to the investor, if it is other than the servicer. As agribusiness finance becomes more sophisticated, *servicing* is routinely handled by software. Inquiries by borrowers are initially directed to a digital phone-answering system, where questions regarding such things as loan balance, remaining term, and interest paid are answered. Loan servicing can also include the administration of the loan when problems are encountered, such as the borrower's inability to make timely payments or failure to provide property hazard insurance or pay property taxes. Any of these events can constitute loan default.

How does the loan process for a short-term loan, such as for working capital needs, differ compared to a long-term, real-estate loan? The seven phases still describe the process but emphasis occurs at different steps. Appraisal of the personal property such as crop or cattle begins immediately after the application is taken, and the processing (phase 2) is on hold until the results of the appraisal are available. Although the underwriting and document phases proceed as in the real-estate loan, recording of any security interest should precede funding. In practice, however, recording can be delayed by the volume of documents at the local recorder's office, so funding often precedes recording and closing.[3] The final phase, servicing, proceeds as with real-estate loans but for a shorter time period. The market for the sale of short-term loans to investors has not developed to the extent that it has for real-estate loans, so servicing is more likely to be on behalf of the original lender.

■ Default and Foreclosure

When a borrower has failed to meet any of the terms of the promissory note, the borrower is in default of the note. For a personal property loan such as farm equipment, the lender already has title to the collateral so the lender only has to take physical possession. This is considered a drastic step by today's commercial and federally sponsored agency lenders, so effort is made to accommodate the borrower. This **lender forbearance**, goodwill efforts on the part of the lender to work with the distressed borrower to avoid possession of the security, may result in a promissory note modification that provides more lenient terms for the borrower such as additional time or a lower interest rate.

Real-estate lenders can periodically experience market downturns when landowners are less able to meet their obligations under the note. During these agribusiness or real-estate slumps, lenders organize a team of employees into "workout" units that specialize in finding ways to assist the borrower in avoiding **foreclosure**, the legal process of recovering the real-estate collateral when the borrower is in default of the promissory note.

For the majority of states where the mortgage is in use, this process is the formal suit of foreclosure. This lawsuit requires filing a petition in superior court in the county where the property is located. Because most court systems are overloaded with civil cases, foreclosure can be a time-consuming process. Therefore, in states where the deed-of-trust system is established, most lenders prefer its usage. If the borrower defaults on the note obligation, the lender enlists the third party or trustee, who files the proper public notice. After an approximate six-month waiting period, the real-estate collateral is sold at public auction to the highest bidder with the lender's loan balance being its bid. Because the deed-of-trust system is less time consuming and less costly, lenders prefer to use it as their security interest in the fifteen states where it is recognized by state law.

If the borrower has defaulted on the note but manages to file voluntary bankruptcy proceedings prior to foreclosure, the borrower may be able to obtain a **stay of foreclosure**, a temporary postponement of the foreclosure proceedings until the court can evaluate overall indebtedness. This stay may delay foreclosure a year or more, adding incentive to the lender's preference for the deed-of-trust system in states where it is available. Ultimately, the lender will either work with the borrower to modify the promissory note or receive the real-estate collateral or its value from the foreclosure process. Fortunately, in years of normal business-cycle expansion, loan defaults leading to foreclosure occur in less than 1% of existing mortgages.

■ Determinants of Interest Rates

We learned in chapter 1 that agribusiness debt capital is accessible to users of funds through either financial intermediaries such as commercial banks, or middlemen, such as loan or securities brokers. Both of these financial market makers operate in the macroeconomy, that is, the national capital markets. Competition to supply funds and to obtain them is keen. To some extent, the economic principles of supply and demand affect market interest rates. All else being equal, the greater the demand for loans given a fixed supply of funds, the higher the interest rate that lenders can charge. The interest rate acts like the "price" of the loan, and, for a given quantity of loanable funds, a strong demand will cause the equilibrium "price" to be higher.

But woven into domestic credit-market history is the behavior of our central bank, the Federal Reserve. Originally created in 1913 by a nervous Congress as a lender of last resort for banks, the Fed has evolved to the role of manager of the nation's money supply, and the national watchdog for inflation. The Fed's policy for the past two decades has been to fight inflation through its **monetary policy**, formal management of the domestic money supply, even at the expense of business-cycle expansion. Inflation is deemed a greater evil than short periods of recession (two or more quarters of productivity decline) that the Fed's tight money policy might bring on.

If the Fed views the current or projected inflation rate as excessive, it raises those interest rates that it can directly control, the **discount rate**, its loan interest rate to banks, and the **Fed Funds rate**, a short-term loan rate between banks for liquidity. This triggers a chain reaction: banks pay more to borrow from the Fed and they charge a higher rate to business and consumer borrowers. Borrowing and lending then slow down, the economy cools off, recession sets in, and inflation subsides. Usually, it takes several quarters for the desired inflation cooling effect to occur.

Separately, lenders and investors consider inflation a definite evil, likely to erode future purchasing power when loan payments are received. So as investors and lenders perceive inflation to be heating up, they independently raise interest rates to hedge their future purchasing power. The upshot is that when inflation is on the rise, the combined effect of the Fed's money-supply tightening and investors and lenders raising their interest rates results in higher interest rates for the agribusiness borrower. Fortunately, the 1990s saw relatively low inflation, keeping long-term rates trending downward much of the decade and into the new millennium.

■ Questions and Problems

2-1. Briefly describe the secondary market for promissory notes.

2-2. In a deed-of-trust state such as California, which party is responsible for sale of the real estate if the borrower defaults on the promissory note? What is the legal term for the borrower in mortgage states?

2-3. Your first home costs two hundred thousand dollars. You put 20% down and finance the rest with a loan from the local branch of the Farm Service Agency. The FSA charges you two points. What do the points cost you in dollars?

2-4. For problem 2-3 above, is it a conforming or jumbo loan?

2-5. Put the following phases for the loan process in proper order: funding, closing, underwriting, application, processing, documents, servicing.

2-6. Suggest two alternatives found in the real-estate industry that are utilized in lieu of foreclosure.

2-7. Using the simple supply-and-demand graph below, show what will happen to interest rates if the Federal Reserve reduces the quantity of money in circulation.

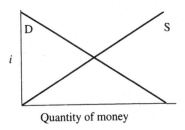

Quantity of money

Figure Q2-7

■ Notes

1. Bill West and Richard Dickinson, *Street Talk in Real Estate* (Unique Publishing, 1987).
2. Fannie Mae, Telephone interview, July 31, 1998.
3. Thomas McGuire, Central Coast Farm Credit, Telephone interview, July 31, 1998.

3
Time Value of Money, Loan Calculations, and Analysis

Many decisions in finance cannot be made without the consideration of interest earned or paid over time. For international currency transactions, individual homeowner financing, and rancher or grower agribusiness financing, an understanding of compound interest and the time value of money is crucial to choosing the correct course of action.

■ The Time Value of Money

The domestic and international credit markets, where the trading of corporate bonds, government securities, and more exotic debt instruments such as mortgage-backed pass-through securities takes place, provide quotations to the press, finance professionals, investors, and other interested parties. For a particular security, these quotations can be in two different formats, either by interest rate or by price. Even a casual observer of credit-market trading will eventually come to the conclusion that for a particular security, if the interest rate has increased, its price will have declined. In another words, market interest rates and debt security prices are inversely related. Why?

When the Dutch West India Company purchased Manhattan Island from Native Americans in 1626 and paid a mere sixty Dutch guilders, or $24, this was considered a "steal." But if the Native Americans had been given the opportunity to invest the money received and earn compound interest, was it such a low price after all? And Seward's Folly, which was purchased on behalf of the United States for a grand total of $7,200,000 in 1867 and ultimately became the State of Alaska, may have been one of the best investments by any government in history. To first understand the proper perspective from which to analyze these issues and then to determine their answers, we begin our study of the time value of money.

■ Compound Interest

A dollar that was invested in a prior period in a typical passbook savings account will earn interest both on the initial principal and on any interest accumulated prior to the current period. In other words, the depositor is earning "interest on interest," assuming the depositor has not withdrawn the interest from prior periods. This is the concept of compound interest. **Compound interest** is interest added to the principal, which from that point forward earns interest too. Most regular savings accounts, including interest-earning checking accounts at depository intermediaries, offer compound interest. From the bank's perspective, this is the "cost" of using the depositor's funds. Although it would be less costly to pay only simple interest, market competition for deposit dollars has forced most intermediaries to provide accounts that offer compound interest.

How is the value of a passbook savings account determined after several periods have elapsed, assuming no withdrawals? This calculation is known as the future value of a sum and is determined using equation (3.1):

$$\text{Future Value of a Sum} = (\text{Present Value}) \times (1 + i)^n \qquad (3.1)$$

in which

Future Value of a Sum is the ending account value (FV),
Present Value is the original principal (PV),

i is the periodic interest rate paid on the account,

n is the number of periods funds are on deposit.

For example, $1,000 invested for four years, earning 6% interest with annual compounding, would be worth $1,262 if left untouched. Equation (3.1) would look like:

$$\text{FV of a Sum} = (\$1,000) \times (1.06)^4 = (\$1,000) \times (1.26247) = \$1,262$$

Invested at 6%, how much would the surviving Manhattan Native Americans have had in 1998, 372 years after the sale of their island?

$$\text{FV of a Sum} = (\$24) \times (1.06)^{372} = \$62,229,042,720!$$

At over $62 billion, perhaps the Native Americans were actually good negotiators.

■ Intraperiod Compounding

Equation (3.1) is useful whenever you want to determine the future value of an investment, assuming a specific interest rate and investment-holding period. Fortunately for depositors, most savings and checking accounts that pay compound interest compound it more frequently than once per year. Quarterly compounding is common for savings deposits, while demand-deposit accounts frequently pay interest compounded monthly.

How does this change equation (3.1)? In the credit markets, n represents years. To account for intraperiod compounding, we modify the equation by dividing the interest rate i by the number of compounding periods, k, that exist per year, and multiplying the n years by k compounding periods per year.

$$\text{Future Value of a Sum} = (\text{Present Value}) \times \left(1 + \frac{i}{k}\right)^{n \times k} \qquad (3.2)$$

where k is the number of compounding periods per year. For our $1,000 investment, held four years, earning 6%, if we assume quarterly compounding, we achieve the following future value:

$$\text{FV of a Sum} = (\$1,000) \times \left(1 + \frac{.06}{4}\right)^{4 \times 4} =$$
$$(\$1,000) \times (1.015)^{16} = \$1,269$$

The additional $7.00 is due to the more frequent compounding, earning interest on interest sooner and more often. So when faced with intraperiod compounding, the rule of thumb to remember is *divide the interest rate* i *by* k *and multiply the number of years* n *by* k, *where* k *represents the number of compounding periods per year.*

How about Seward's Folly? Did the United States get a good deal on that purchase 130-plus years ago? What would the purchase price for Alaska be in 1998, assuming it increased at 6% per year with quarterly compounding?

$$\text{FV of a Sum} = (\$7,200,000) \times (1.015)^{524} = \$17,601,052,366$$

Since the mineral reserves alone are worth that price, it looks like Seward did us a favor!

Equations (3.1) and (3.2) contain the parenthetic expressions that compute an interest factor, given the interest rate i and the term n. Prior to the convenience of handheld calculators and personal computers, the calculations we have been doing were done longhand with the aid of tables. These tables included future value of a dollar factors with various interest-rate and term combinations. For your convenience in working problems, we have included the factors in tables in the appendix. But to free yourself from the need to look up any factor, just use the exponential key on your calculator. Problems can also be solved using a financial calculator, in the time-value menu or row of keys. For the balance of this chapter, we will provide example solutions that utilize either the factor tables or the calculator exponential-key method.

■ The Process of Discounting

You have probably heard the adage "a dollar today is worth more than a dollar in the future." This agrees with our intuition, if only because a dollar today is like a bird in hand; it's a certainty. There is always some risk that we will not receive the dollar tomorrow, whether it is the risk of borrower default, some market event like a bank run or crop failure, or possibly our own failure to show up to collect. But investors also consider other issues. Purchasing power is key to market interest rates, as we learned in chapter 2. Investors "discount" future dollars partly because of anticipated lost purchasing power as a result of inflation. A third reason the future dollar is less valuable is opportunity cost. Giving up a dollar today in order to receive it in the future limits your options. You can no longer invest today's dollar in something else; it's tied up in the transaction to receive the future dollar.

When investors attempt to estimate the value today of a future dollar (or other currency for that matter) or cash flow, they employ the process of discounting. **Discounting** is the compounding of interest in reverse for a future value to determine its present value. Mathematically, equation (3.1) is modified as given below in equation (3.3):

$$\text{Present Value} = (\text{Future Value}) \times (1 + i)^{-n} \tag{3.3}$$

For example, suppose Manhattan Island is worth one hundred billion dollars in 1998. What should the Trading Company have paid for it in 1626 if their "discount rate," the appropriate interest rate for the valuation, given risk, infla-tion, and opportunity cost, equaled 6%?

$$PV = (\$100,000,000,000) \times (1.06)^{-372} = \$38.57$$

Perhaps the Indians were underpaid after all!

Consider a modern example involving your state's lottery. Suppose you win the big jackpot, worth either ten million dollars today, or twenty million dollars in ten years. Assuming taxes remain unchanged during that time and that your chosen discount rate is 8%, the better choice can be determined one of two ways. We can either compound the ten million dollars present value to its future value in ten years, or discount the deferred payment for ten years at an 8% dis-count rate. The discounting method is given below:

$$PV = (\$20,000,000) \times (1.08)^{-10} = \$9,263,870$$

We chose a discount rate high enough to make the jackpot awarded today more attractive. But note that if our discount rate were merely 1% less, the deferred cash flow would be chosen, worth $166,986 more than today's $10,000,000 jackpot.

Present-value calculations are useful in many finance contexts. For this text, we will frequently rely on equation (3.3) or its intraperiod compounding variation (3.4), given below:

$$\text{Present Value} = (\text{Future Value}) \times \left(1 + \frac{i}{k} \right)^{-(n \times k)} \qquad (3.4)$$

where k is the number of compoundings per period or year. Suppose for our lot-tery example compounding was semiannual, that is, twice per year.

$$PV = (\$20,000,000) \times (1.04)^{-20} = \$9,127,739$$

This is $131,136 less than our original solution. Why? The discounting process takes place sooner and twice as often. More frequent discounting always low-ers the present value of the future cash flow, ceteris paribus.

■ Annuities

An **annuity** is a series of equal, periodic cash flows over a finite life. There exist two types of annuities in modern finance, ordinary and annuity due. The **ordinary annuity** has cash flows at the end of each period. Loans are typically repaid in the form of an ordinary annuity. **Annuities due** have cash flows at the beginning of each period. They are common in the insurance and retirement investment fields, where the initial contribution is made at the start of the pro-gram. Our primary focus will be on the loan market and the ordinary annuity.

How do repeat cash flows, that is, an annuity, affect our effort to value the cash flows? We can find the future or present value the "long way" by repeatedly using either equation (3.1) or (3.3). But the mathematics has long been available to shorten this process. For the future value of an ordinary annuity:

$$\frac{\text{Future value}}{\text{of an annuity}} = \left(\frac{\text{Periodic}}{\text{cash flow}}\right) \times \left(\frac{(1+i)^n - 1}{i}\right) \qquad (3.5)$$

For example, suppose you invest $1,000 per year at the end of each year (ordinary annuity) toward your future retirement, planned forty years from today, and the bank guarantees you 8%. The investment's ending or future value would be found as follows:

$$\frac{\text{Future value}}{\text{of an annuity}} = (\$1,000) \times \left(\frac{(1.08)^{40} - 1}{.08}\right) = \$259,057$$

of which $219,057 is compound interest. Had we used the appendix, we would look up the future value of an ordinary annuity interest factor, which equals 259.057, and multiply it by the cash flow per year of $1,000, to get the same answer as before. Either way, the power of compounding is clearly illustrated. In this case, it pays to begin retirement investing early.

Because some investments such as whole life insurance policies are paid at the start of the period, let's examine how the timing of the cash flows affects the future value. A simple case would be that of a two-year annuity, earning 10%, when we want to know its future or ending value after two years. For the ordinary annuity, from equation (3.5) we determine the value to be $2,100. But for the beginning-of-period annuity, the annuity due, the future value becomes $2,310. The additional $210 is due to compounding. The initial payment earns interest for a full two years, while the second payment earns one full year of interest. Mathematically,

$$\frac{\text{Future value of}}{\text{an annuity due}} = \left(\frac{\text{Periodic}}{\text{cash flow}}\right) \times \left(\frac{(1+i)^{n+1} - 1}{i} - 1\right) \qquad (3.6)$$

■ Present Value of an Annuity

Most state lotteries pay their jackpots over several years. Our lottery problem above was unrealistic in one sense: you would most likely receive the ten million dollars over twenty years in installments of five hundred thousand dollars each. Now suppose you could sell the future jackpot installments to an investor who believed the appropriate discount rate is 10%, considering his risk, inflation, and opportunity cost. How much would you receive today?

Although the initial payment would most likely be at the beginning of the twenty years, let's assume all payments fall at the end of each period, an ordinary annuity. To determine how much the investor would pay you for your future winnings, we need equation (3.7):

$$\text{Present value of an annuity} = \left(\frac{\text{Periodic}}{\text{cash flow}}\right) \times \left(\frac{1 - (1 + i)^{-n}}{i}\right) \quad (3.7)$$

For our example,

$$\text{Present value of an annuity} = (\$500,000) \times \left(\frac{1 - (1.10)^{-20}}{.10}\right)$$

$$= (\$500,000)(8.51356) = \$4,256,782$$

which reveals the strength of discounting. If we sell the future jackpot cash flows to the investor at the above price, he earns $5,743,218 in interest!

But what if the first installment were paid immediately? Intuition tells us the discounted jackpot would be worth more than before. In other words, an annuity due is worth more than an ordinary annuity, ceteris paribus. Equation (3.8) is utilized to discount annuities due.

$$\text{Present value of an annuity due} = \left(\frac{\text{Periodic}}{\text{cash flow}}\right) \times \left(\frac{1 - (1 + i)^{-(n-1)}}{i} + 1\right) \quad (3.8)$$

For our lottery jackpot, paid over twenty years, at $500,000 per year, beginning now and received at the beginnings of the subsequent years, the present value is $4,682,460. The fact that the initial payment is received at the very start and all other payments are received a year sooner is worth an additional $425,678. At least for the initial payment, there is little uncertainty, opportunity cost, or loss of purchasing power to reduce its value.

For annuities, intraperiod compounding is handled the same way that it is for one-time cash flows. Merely divide the interest rate i by k, the number of compoundings per period, and multiply n periods by k. Returning to our retirement-investment example, suppose that instead of $1,000 per year for forty years, you set aside $250 at the end of every three months. For forty years, with quarterly compounding at an 8% annual rate, the future or ending value of your investment would be:

$$\text{Future value of an annuity} = (\$250) \times \left(\frac{(1.02)^{40 \times 4} - 1}{.02}\right) = \$283,624$$

The extra $25,567 is from the intraperiod compounding, earning interest on interest sooner and more often. Suppose you were using the annuity tables in the appendix. How would you adjust for the quarterly compounding? The tables would need to include 2% for 160 periods, but our tables only go out 45 periods. Therefore, it is much easier in the long run to learn to solve problems using the exponential-key method.

■ Basic Loan Calculations

Consider our lottery problem one more time. Suppose that instead of selling the future jackpot cash flows to an investor, you need $4,250,000 to open your microbrewery and have found a lender willing to advance you the funds with your future lottery cash flows as collateral. Like the investor, the lender requires 10% interest with annual payments. How large would these payments be?

For a fully amortizing loan, introduced in chapter 2, we know that each payment is equal; that is, it is in the form of an annuity. Interest is paid by the borrower and earned by the lender each period, and payments are made at the end of each period, making the payments an ordinary annuity. The original loan amount is received in the present, making this an ordinary annuity problem. We know from equation (3.7) that the present value of the annuity equals the cash flows times the annuity factor. So, from simple algebra, we find:

$$\frac{\text{Periodic}}{\text{cash flow}} = \frac{\text{Loan}}{\text{payment}} = \frac{\text{Present value of an annuity}}{\left(\dfrac{1 - (1 + i)^{-n}}{i}\right)} \qquad (3.9)$$

and our payments on the $4,250,000 loan will be:

$$\frac{\text{Loan}}{\text{payment}} = \frac{\$4,250,000}{\left(\dfrac{1 - (1.10)^{-20}}{.10}\right)} = \$499,203$$

which is close to our $500,000 annual jackpot cash flow.

For any amortizing loan, the principle is the same. You merely divide the loan amount by the present value of an ordinary annuity factor. Since the loan is fully amortizing, the principal balance will be $0 at the end of its term, twenty years in this example.

An alternative method to solving for the amortized loan payment is again based on the fact that division by a fraction is the same as multiplication by the reciprocal of that fraction. We can take the reciprocal of the present value of an annuity factor found in formula 3.9 and *multiply* it by the original loan principal (the present value of the annuity) instead of dividing. This process is shown for the same example as above:

$$\frac{\text{Loan}}{\text{payment}} = \left(\begin{array}{c}\text{Present value} \\ \text{of an annuity}\end{array}\right) \times \left(\frac{i}{1 - (1 + i)^{-n}}\right)$$

$$\frac{\text{Loan}}{\text{payment}} = (\$4,250,000) \times \left(\frac{.10}{1 - (1.10)^{-20}}\right)$$

$$= (\$4,250,000) \times (0.1174596) = \$499,203$$

Notice that we obtain the same loan payment. The factor that results from finding the reciprocal of the present value of the annuity factor is called the amortization factor. A table of these factors for whole annual interest rates for terms up to forty-five years can be found in the appendix.

■ Building an Amortization Schedule

As illustrated in table 2.2, the schedule can be provided in six columns. Once you have calculated the fully amortizing loan payment using equation (3.9), then begin by starting a column for the periods or years. If the loan has monthly repayment, as most do, then the first column would be for each month. The second column, beginning balance, is the balance at the start of that period or month. Next, calculate the interest for the period by multiplying the interest rate times the beginning balance. Place the product in column 4, then subtract the interest paid for that month from the total payment (listed in column 3). The difference is the principal paid for the current period and is placed in column 5. Finally, subtract the current principal paid (column 5) from the period's beginning balance (column 2), and place the difference in column 6, ending balance. The ending balance for period one becomes the beginning balance for period two. The first two periods for our microbrewery loan are shown in table 3.1.

Table 3.1. First two years of microbrewery loan amortization schedule

(1)	(2)	(3)	(4)	(5)	(6)
Year	Beginning balance	Payment	Interest	Principal	Ending balance
1	$4,250,000	$499,203	$425,000	$74,203	$4,175,797
2	4,175,797	499,203	417,580	81,623	4,094,174

If you were to carry out the schedule through the twentieth year, you would find the ending balance equaling approximately $0. Amortization schedules are an algorithmic process but can easily be written as a spreadsheet program. In addition, financial calculators have built-in amortization routines. The key point is that a correctly calculated loan payment using equation (3.9) will amortize the loan principal over its term, reducing the balance to $0 with the final payment.

■ Loan Balance

Because the loan payment is the quotient of the loan amount and the present value of an annuity factor, the loan balance is the *product* of the loan payment and the present value of an annuity factor. If you use the initial present value of an annuity factor in equation (3.10), the product is the original loan amount. To find the loan balance of a fully amortizing loan at any time during the life of the loan, change the present value of an annuity factor to the number of periods remaining. That is, change the value of n to indicate how many months, quarters, or years are left in the loan term. Equation (3.10) shows how the loan balance can be calculated from the periodic payment:

$$\frac{\text{Loan}}{\text{balance}} = \left(\frac{\text{Loan}}{\text{payment}}\right) \times \left(\frac{1 - (1 + i)^{-n}}{i}\right) \tag{3.10}$$

For the lottery-backed microbrewery loan, the balance after five years would be (fifteen years remaining):

$$\frac{\text{Loan balance}}{\text{at 5 years}} = (\$499,203) \times \left(\frac{1 - (1.10)^{-15}}{.10}\right) = \$3,796,978$$

This end-of-period balance can be checked by continuing the amortization schedule through the fifth year. Clearly, equation (3.10) is useful in avoiding tedious calculations to determine the loan balance.

A related, tax-driven question is what interest was paid for that fifth year? Since in agribusiness we need after-tax information, a convenient formula to determine interest paid over any given period is given in equation (3.11) below:

$$\frac{\text{Interest paid}}{\text{within a period}} = \frac{\text{Total payments}}{\text{within the period}} - \frac{\text{Change in}}{\text{loan balance}} \tag{3.11}$$

To solve for our fifth year, we need the loan balance at the end of year four in addition to the balance at end of year five.

$$\frac{\text{Loan balance}}{\text{at 4 years}} = (\$499,203) \times \left(\frac{1 - (1.10)^{-16}}{.10}\right) = \$3,905,622$$

Then our solution follows:

$$\frac{\text{Interest paid}}{\text{within year 5}} = [(\$499,203) - (\$3,905,622 - \$3,796,978)] = \$390,559$$

In prose, the difference of the two loan balances is the amount of the fifth-year payment that went to principal. The rest went to interest. This interest expense is deductible and is reported to the Internal Revenue Service.

Is there an easy way to determine the total interest paid over the life of the loan? Equation (3.11) provides the answer.

$$\frac{\text{Interest paid}}{\text{within 20 years}} = [(20 \times \$499,203) - \$4,250,000] = \$5,734,060$$

where the $4,250,000 is the total principal paid over the life of the loan and (20 $\times$ $499,203) reflects the total payments.

Determining the Annual Percentage Rate

In chapter 2, we discussed truth-in-lending requirements of Regulation Z. The Federal Reserve Bank, in cooperation with other federal regulators, has created disclosure requirements for conventional lenders doing federally related loans, including estimating and providing the true annual percentage rate, or APR. One common example is that of financing your new truck purchase. If the vehicle is for personal use, the APR must be disclosed by conventional lenders. In effect, the APR is the true or actual interest rate for the loan, given any fees charged. Assume you desire a new three-quarter-ton pickup truck for equipment hauling and recreational use. The cost is $25,000, including tax and license. You decide to put 20% down and finance the rest through your local community bank. The bank charges you two points to provide this loan at 6%, for five years, with monthly payments. Because of the points, the lender must disclose the APR to you prior to your signing the promissory note. To determine the APR, first we calculate the payment using equation (3.9):

$$\frac{\text{Loan}}{\text{payment}} = \frac{\substack{\text{Present value} \\ \text{of an annuity}}}{\left(\dfrac{1 - (1 + i)^{-n}}{i}\right)} = \frac{\$20,000}{\left(\dfrac{1 - (1.005)^{-60}}{.005}\right)} = \$386.66/\text{month}$$

Next, we use equation (3.10), the loan-balance equation, but adjust the initial loan amount for the points paid; in other words, subtract the two points from the loan amount.

$$\frac{\text{Loan}}{\text{balance}} = \left(\frac{\text{Loan}}{\text{payment}}\right) \times \left(\frac{1 - (1 + i)^{-n}}{i}\right) = (\$20,000 - \$400) =$$

$$\$19,600 = (386.66) \times \left(\frac{1 - \left(1 + \dfrac{i}{12}\right)^{-60}}{\dfrac{i}{12}}\right)$$

where we then use trial and error to estimate i. What you, the borrower, experience, is making the payment as agreed to by the terms of the loan, but you really only receive $19,600 from the lender. You had to pay $400 to obtain the $20,000 loan, so your net loan amount is the difference.

The trial and error estimate of i at first seems a difficult task, but after some practice, your knowledge of appropriate interest rates will improve and the process will become much easier. For our truck loan, an estimate of the APR within .125% (one-eighth percent) will be sufficiently accurate. Let's try 7% for our first estimate. Using 3.10:

$$\text{Loan balance} = \$386.66 \times \left(\frac{1 - (1.0058)^{-60}}{.0058} \right) =$$

$$(\$386.66) \times (50.5020) = \$19,527$$

This value is slightly too small; that is, it is less than $19,600. Since interest rates and values are inversely related, let's try 6.875% for our second attempt.

$$\text{Loan balance} = \$386.66 \times \left(\frac{1 - (1.0057)^{-60}}{.0057} \right) =$$

$$(\$386.66) \times (50.6527) = \$19,585$$

This tells us that the closest APR interest rate is 6.875%. So in just two trials, we solved for the APR estimate.

Several shortcuts are worth noting in APR estimation. First, the APR will always be higher than the loan's stated interest rate if points are paid. For shorter-term loans, ceteris paribus, the APR will be higher than for longer-term loans. And for real-estate mortgages, with a term of thirty years, a good rule of thumb is that two points raise the APR 1/4% above the loan's stated interest rate.

Regulation Z protects farm purchasers as long as the real estate is used for the owner's residence. The following real-estate example is typical of a first-time rural home purchase:

Purchase price	$250,000
Down payment	20%
Interest rate	6%
Term	30 years
Payments	monthly
Points	2

We first calculate the monthly payment using equation (3.9):

$$\frac{\text{Loan}}{\text{payment}} = \frac{\text{Present value of an annuity}}{\left(\dfrac{1 - (1 + i)^{-n}}{i} \right)} = \frac{\$200,000}{\left(\dfrac{1 - (1.005)^{-360}}{i} \right)} =$$

$$\$1,199.10/\text{month}$$

Next we adjust the loan amount for the two points, or $4,000. So the APR trial-and-error equation is:

$$\$196{,}000 = (\$1{,}199.10) \times \left(\frac{1 - \left(1 + \dfrac{i}{12}\right)^{-360}}{\dfrac{i}{12}} \right)$$

and we try (using the rule of thumb) 6.25% for our interest rate. Then the net loan amount is \$194,732:

$$(\$1{,}199.10) \times \left(\frac{1 - (1.0052)^{-360}}{.0052} \right) = (\$1{,}199.10) \times (162.41) = \$194{,}732$$

We have calculated a net loan balance that is too low, indicating that our trial interest rate is too high. Let's try 6.125%.

$$\text{Loan balance} = (\$1{,}199.10) \times \left(\frac{1 - (1.0051)^{-360}}{.0051} \right) = \$197{,}330$$

This is too high a net loan balance. The actual net loan balance of \$196,000 is about halfway between the two values, slightly closer to \$194,732. So the APR is closest to 6.25%. From a financial calculator, the precise APR is 6.1895%, again closer to 6.25%. With practice, you will be able to estimate APRs with two or three trials. If you prefer, buy yourself a powerful financial calculator instead!

■ Refinance Analysis

For more than two decades, credit-market interest rates have been rather volatile, fluctuating from day to day, and changing dramatically over the business cycle. The late 1990s was a fairly stable period, but with downward-trending long-term interest rates. Many consumers and agribusinesses have undergone refinancing of their real-estate debt to capture these lower rates. For consumers, one "rule of thumb" has gained popularity: *consider refinancing if current market interest rates are at least 2% below your existing fixed-interest-rate mortgage.* Unfortunately, this has not always been sage advice.

The proper perspective for refinancing is to weigh the discounted cash-flow savings of the new, lower payment against the cost of the transaction. Once again we must discount the future cash-flow savings because they occur in the future and suffer all the consequences of discounting—lost purchasing power, opportunity cost, and uncertainty. Of course, some refinance transactions cover the initial cost of the new loan in the loan amount itself. If the new payments are lower, it seems a straightforward conclusion that the refinance is a good move. But there is usually a catch. The new loan probably extends your term; that is, payments must now be made for a longer time period. By refinancing you have locked in the obligation of making the new payments many years into the future, when for at least some years you would have had *no* payment if you had kept the original mortgage. So each refinance proposal can be just complicated enough to invalidate "rules of thumb" and may require careful analysis. We will analyze several examples.

First, suppose five years ago you had borrowed $200,000 to finance your fruit-packing facility with terms of 9%, thirty years, and monthly payments. Today, the market interest rate is 8%, you can get a twenty-five-year loan for two points plus $1,000 in other fees, and the bank will let you finance the cost of the loan. This means you will have no out-of-pocket expenses. Since the original loan is five years old, and the new loan would be for twenty-five years, you would not extend your payment by refinancing. Is this "refi" proposal a good idea? To solve it, we merely need to compare the old and the new monthly payments.

Using equation (3.9), we determine the original monthly payment to be $1,609. But the loan is five years old, so its current loan balance, from equation (3.10), is $191,760. The new loan amount is found algebraically as follows:

$$\text{New loan amount} = \frac{(\$191{,}760 + \$1{,}000)}{1 - .02 \text{ points}} = \$196{,}695 \cong \$196{,}700$$

If we borrow $196,700, subtract $1,000 to pay miscellaneous fees and another $3,934 for points, we'll still have enough to pay off the old loan balance ($191,766) with $4.00 to spare. Now all we need to do is calculate the new payment at 8% for twenty-five years using equation (3.9).

$$\text{Loan payment} = \frac{\text{Present value of an annuity}}{\left(\dfrac{1 - (1 + i)^{-n}}{i}\right)} = \frac{\$196{,}700}{\left(\dfrac{1 - (1.0067)^{-300}}{.0067}\right)} = \$1{,}518$$

Payments of $1,518 are $91 per month lower than the original payment. In this simple case, where the term is not extended and all costs are financed, clearly it pays to proceed with the refinance.

Suppose we want to refinance the same original loan, but the new lender requires the loan costs to be paid out of pocket. Does the refi still make sense? If costs are paid out of pocket, then we only need to borrow $191,766 to pay off the original loan. Our new payment at 8%, twenty-five years, monthly payments, would be $1,480. This is a savings of $129 per month, but at a cost of two points plus the $1,000 lender fee. We must decide if the $129 per month savings for twenty-five years is worth the total cost of $1,000 plus $3,835 in points, or $4,835.

One simple way to evaluate the savings versus the cost is to determine the breakeven in months.

$$\frac{\text{Breakeven}}{\text{in months}} = \frac{\text{Loan costs}}{\text{Payment savings}} = \frac{\$4,835}{\$129/\text{month}} = 38 \text{ months}$$

This implies you would recover your initial costs after making the reduced payment for about three years, after which you would continue to save $129 every month for the remaining twenty-two years.

But what about the time value of money? Those future monthly savings should be discounted and be weighed against the initial cost of refinancing. Let's solve for the yield on our $4,835 investment. We must utilize equation (3.10) (our loan balance and APR formula) and trial and error to find the yield, using 24% for our first trial.

$$\frac{\text{Present value}}{\text{of an annuity}} = (\$129) \times \left(\frac{1 - (1.02)^{-300}}{.02} \right) = \$6,433$$

This answer of $6,433 is too high a value—it is larger than $4,835. Trying 40%, the present value of the monthly savings is $3,870. Therefore, the yield on our initial cost invested in the refinance lies between 24% and 40%. Several more trials determine the yield to be closest to 32%. This implies that unless there is a competing investment that will yield more than 32% for our $4,835 loan-cost investment, we should proceed to refinance.

As noted above, sometimes the analysis is not this straightforward, due to timing differences. Suppose we still were offered the 8% interest rate, but the new loan would *also* be a thirty-year loan. This means we would have a payment savings during the first twenty-five years, but have to continue to make payments beyond the twenty-fifth year for the new loan, when we would have had no payment under the original loan. So, we trade off the loan costs at the beginning plus five additional years of loan payments at the end for twenty-five years of payment savings. This analysis is more easily solved on a financial calculator with a cash flow menu. As a general rule, however, if the payment savings occur over a much longer period than the additional new loan payments at the end, then the refinance should proceed. Using a financial calculator, we determine the investor only needs to earn a yield of .54% per year to profit from this refinance.

The question of whether to refinance can usually be solved using one of the above approaches. This issue is most important during periods of falling interest rates such as we experienced prior to the turn of the century. As an agribusiness manager, you should be on the lookout for refinance opportunities.

■ Cash Budgeting

Working-, or operating-, capital loans are usually provided from a commercial bank, the local Farm Credit System association, the Farm Service Agency, or a private lender. Banks and the Farm Credit branches offer this financing in the form of lines of credit. The agribusiness is approved to borrow any amount within the maximum loan amount for, say, an annual period, at the end of which the total must be paid off. Because the lender usually extends the line of credit for a future period, this can be viewed as a "revolving" line of credit or revolving debt. But revolving debt also includes credit card financing, which can remain outstanding indefinitely. Most of us have been besieged by credit card companies, and it is in their best interest to encourage a large balance, near the credit line's maximum. This maximizes the interest income earned by the credit card company.

But for agribusiness lines of credit, the lender wants to see the balance brought to zero for at least one month out of the year. The zero balance demonstrates the borrower's ability to manage the credit, and the successful end to the borrower's business year.

In applying for a working-capital line of credit, you are typically asked to provide a one-year cash-budget forecast. There are a variety of formats for cash budgets; we have provided an eight-line or eight-category format. Table 3.2 shows the basic components of the eight-line cash budget.

Table 3.2. Cash budget format

Month	January	February	March
Cash receipts			
Less: cash payments			
Net cash flow			
Plus beginning cash			
Cash without borrowing			
Current borrowing/surplus			
Cumulative borrowing			
Ending cash balance			

Cash receipts and cash payments are the given month's cash flows. Sales for cash would be an inflow, as would the collection of an account receivable from a prior month's sale. Payment of rent (by check) for leased equipment would be an outflow, as would paying wages and salaries for employees. A simple example follows.

Jensen Feed & Supply has both cash sales and credit sales. All credit sales are collected one month following the month of sale. Sales for December, January, February, and March are forecasted to be $28,000, $42,000, $45,000, and $26,000, respectively, of which 50% are for cash and 50% are on credit. Cash payments (expenses) for January, February, and March are forecasted to be $41,000, $49,000, and $16,000, respectively. January's beginning cash balance is $5,000, which is Jensen's minimum acceptable balance. Complete the cash budget for January through March (ignore interest expense).

The solution first requires computing each month's cash receipts. For January, receipts total half of the current month's sales plus half of December's, which are collected in January. So, $21,000 + $14,000 equals total receipts of $35,000 for January. Similarly, receipts for February and March are $43,500 and $35,500, respectively. These figures should be placed in the appropriate cells of the schedule. Since the cash payments were given in this simple example, those figures are placed in the cash-payments cells. The difference of each month's receipts and payments becomes the net cash flow for the month and is entered in the appropriate cell. At this point, the incomplete cash budget would appear as in table 3.3.

Table 3.3. Partial cash budget

Month	January	February	March
Cash receipts	$35,000	$43,500	$35,500
Less: cash payments	41,000	49,000	16,000
Net cash flow	−6,000	−5,500	19,500

The next step in the budget's preparation is to add the month's net cash flow to the beginning cash balance, to get the total cash without borrowing. In months of surplus cash flow, there is likely to be no borrowing, or at least some repayment of the outstanding cumulative borrowing. In months like our budget's January and February, we must resort to current borrowing to bring our ending cash balance up to the minimum required $5,000 to begin the following month. For January, these calculations appear in table 3.4.

Now we must place January's ending cash balance into the cell for February's beginning cash balance. February starts off with the minimum $5,000 balance, but because of the *negative* $5,500 cash flow, we must borrow the additional $5,500 in February. Our end-of-February cumulative borrowing is $11,500. We end February with the minimum $5,000 cash balance once again, and so start off March with that beginning balance.

But March is a month of $19,500 positive net cash flow. Prior to paying off the cumulative loan balance of $11,500, we have a "cash without borrowing" balance of $24,500. It is prudent to pay off as much of the line of credit as possible in months of cash-flow surplus, minimizing the cost of interest expense.

Table 3.4. January complete cash budget

Month	January	February	March
Cash receipts	$35,000	$43,500	$35,500
Less: cash payments	41,000	49,000	16,000
Net cash flow	−6,000	−5,500	19,500
Plus beginning cash	5,000		
Cash without borrowing	−1,000		
Current borrowing/surplus	6,000		
Cumulative borrowing	6,000		
Ending cash balance	5,000		

So the current surplus is $19,500 and $11,500 of it is used to pay off the cumulative loan balance. The March ending cash balance becomes $13,000, as provided in table 3.5.

Table 3.5. Completed cash budget

Month	January	February	March
Cash receipts	$35,000	$43,500	$35,500
Less: cash payments	41,000	49,000	16,000
Net cash flow	−6,000	−5,500	19,500
Plus beginning cash	5,000	5,000	5,000
Cash without borrowing	−1,000	−500	24,500
Current borrowing/surplus	6,000	5,500	Surplus 19,500
Cumulative borrowing	6,000	11,500	0
Ending cash balance	5,000	5,000	13,000

The Jensen example ignores many aspects of cash budgeting, but it does reveal several key concepts. First, agribusinesses must carry a minimum cash balance each month, most likely in the business checking account. This minimum allows the firm to pay immediate invoices without necessarily receiving cash inflows on that day. This minimum cash balance is set after management has observed the cash account over time, and may be raised as the business grows.

Second, surplus cash, the result of positive net cash flows in a given month, should first be used to pay down or pay off the cumulative loan balance. Surplus cash is that amount in cash without borrowing that exceeds the minimum cash balance. For Jensen, the March cash without borrowing balance of $24,500 exceeded the minimum cash balance of $5,000 by $19,500.

Finally, and perhaps most important, the working-capital line of credit needs to be large enough to exceed the highest month's forecasted cumulative borrowing. For these three months, since the highest cumulative borrowing occurred in February, at a figure of $11,500, a $12,000 line of credit may be adequate, providing no other month of the year requires a higher cumulative borrowing figure than February.

How can we improve on the above eight-line cash-budget format? A variation that has become popular with agribusiness borrowers is provided in table 3.6. Note that the beginning cash is entered first, followed by the cash receipts and cash payments. This format blurs the meaning of net cash flow, but has the advantage of detailing the line of credit status more clearly. Both formats, the original eight-line cash budget and the modified sixteen-line budget, conclude each month with an ending cash balance, and carry this figure over to the start of the following month as the beginning cash balance.

Table 3.6. Modified sixteen-line cash budget

Month	January	February	March
Beginning cash			
Cash receipts			
Cash sales			
Receivables			
Total cash receipts			
Cash payments			
Wages and salaries			
Interest expense			
Other cash payments			
Total cash payments			
Net cash flow			
Loan status			
Current borrowing			
Current repayment			
Cumulative balance			
Ending cash balance			

This format also details receipts and disbursements, particularly interest expense. Our Jensen Feed & Supply example follows in table 3.7 using the modified format.

With the additional detail, the agribusiness and the prospective lender can more clearly see the need for the line of credit, and how it will be repaid. Of course, this modified format can be augmented by adding as many line-item rows as necessary in the categories of cash receipts and cash payments. Otherwise the format remains unchanged.

Table 3.7. Jensen sixteen-line cash budget

Month	January	February	March
Beginning cash	$5,000	$5,000	$5,000
Cash receipts			
Cash sales	21,000	22,500	13,000
Receivables	14,000	21,000	22,500
Total cash receipts	35,000	43,500	35,500
Cash payments			
Wages and salaries	10,000	10,000	10,000
Interest expense	0	60	115
Other cash payments	31,000	38,940	5,885
Total cash payments	41,000	49,000	16,000
Net cash flow	−1,000	−500	24,500
Loan status			
Current borrowing	6,000	5,500	0
Current repayment	0	0	11,500
Cumulative balance	6,000	11,500	0
Ending cash balance	5,000	5,000	13,000

Which format should you choose? When you apply for a working-capital line-of-credit loan with a bank or Farm Credit branch, the lender will probably provide you with its preferred format as a preprinted form. If not, the more detail you can provide the loan officer, the better. It shows you have done your "homework" and increases the lender's confidence in your business skills.

■ Graphing Loans

Now that we have the background to do loan calculations, cash budget forecasts, and amortization schedules, is there an easy way to depict different types of loans graphically? That is, will a graph or set of graphs convey the repayment characteristics of a loan? Since a picture is worth, well, a few hundred words anyway, let's conclude our loan-analysis chapter with several simple graphical depictions of agribusiness loans.

Three graphs are sufficient to depict any conventional loan. All three utilize time as the horizontal (x) axis. The first graph depicts principal, or loan balance, on the vertical (y) axis. The second has the dollar payment on the vertical, and the third indicates the interest rate on the vertical axis. Take a fully amortizing twenty-five-year farm loan from the rural community bank. The loan is fully amortizing, payments are fixed for the entire twenty-five years, and the interest rate is also fixed. Without providing quantitative information, this fully amortizing loan would appear as shown in figure 3.1.

For typical commercial term loans to agribusiness suppliers such as the Jensen Feed & Supply example, these loans usually are partially amortizing with a balloon payment, as depicted in figure 3.2.

Are there other common agribusiness loan variations? Lines of credit certainly graph differently than amortizing loans—as noted for a five-year credit line in figure 3.3. And longer-term adjustable- or variable-interest rate loans have their own graphical characteristics—see figure 3.4.

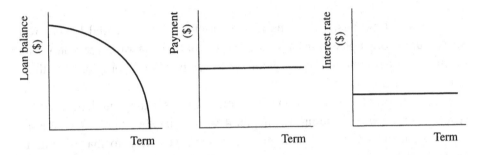

Figure 3.1. Fully amortizing loan.

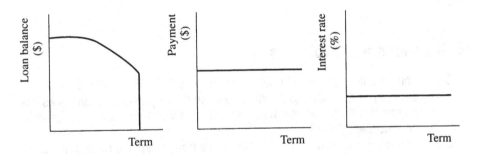

Figure 3.2. Partially amortizing with balloon.

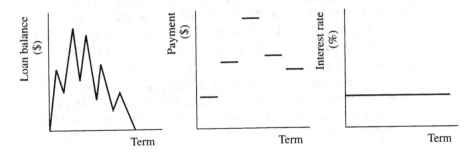

Figure 3.3. Line of credit with annual repayment clause.

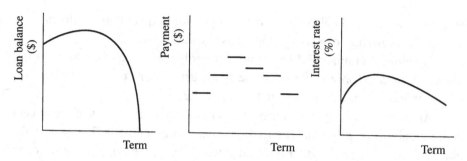

Figure 3.4.   Adjustable interest rate long-term loan.

For most agribusinesses, the above varieties of commercial loans cover the gamut. Another possibility for commercial loans is that of interest only with a balloon. Try graphing that one yourself to see how well you grasp the three graph concepts.

At this point, we have done little more than set the foundation for the study of agribusiness lending. In chapter 4 we will tackle a key time-value-of-money concept and capital budgeting, and then we return to loans in future chapters when we examine the underwriting procedures of agribusiness lenders. For now, you should be able to do any important loan calculation, such as estimating your future student loan payments or, perhaps more exciting, your next new car payment!

■ Questions and Problems

3-1.   You inherit a section of farmland on January 1, 2005, when it is worth $1,000 per acre. Due to nearby suburban development, it appreciates 5% a year. You decide to sell the section on January 1, 2015. How much do you receive in total?

3-2.   When you turn sixty-five (forty-five years from now), you hope to have saved one million dollars for your retirement. If inflation averages 3% between now and then, what will be the purchasing power of your savings in today's dollars?

3-3.   Kraft Foods executives have fifteen thousand dollars placed in their 401K retirement program at the end of each year as part of their compensation. John Trump has worked for Kraft for twenty-five years and his 401K plan has had an average annual return of 10% per year.

   a.   What is the total value of Trump's 401K account?

   b.   Trump retires at the end of his twenty-fifth year and begins withdrawing from the account. He expects to live another twenty years. If the account continues to earn 10% while funds are still on deposit, how much will each annual withdrawal (as an ordinary annuity) be?

c. Trump wants to leave one million dollars to his nephew when he dies in twenty years. If the account still earns 10%, how much will he be able to withdraw each year and still leave the desired inheritance in twenty years, assuming he dies on schedule?

3-4. You've won the lottery and will receive ten thousand dollars per year, paid semiannually, for the next fifty years. If the appropriate discount rate is 10%, what is the present value of your winnings?

3-5. It's finally time to buy that new SUV you've been wanting. The price is thirty-five thousand dollars, you put five thousand dollars down and finance the rest at 6% interest over six years with monthly payments.

a. How much is each monthly payment?
b. How much interest did you pay the second year?
c. What is the total interest paid over the entire six years of the loan?

3-6. Prepare a six-column amortization schedule for the car loan in problem 3-5 and complete the first three months amortization.

3-7. Your first home purchase is financed by a conventional lender, which must follow Regulation Z guidelines. The home purchase price is $250,000, you put 20% down, and finance the balance at 9% for twenty-five years with monthly payments. The lender charges 1.5 points.

a. How much is each monthly payment?
b. What is the dollar amount of the points?
c. To the nearest one-tenth of a percent, what is the loan's APR?

3-8. Some private farm-mortgage loans are partially amortizing with balloon payments. Graph the principal, payment, and interest rate for a balloon mortgage amortizing over twenty-five years but all due at the end of the tenth year.

3-9. Refer back to problem 3-7. After making payments for five years, interest rates decline due to loose monetary policy, and you can refinance the balance of your mortgage at 7% for twenty years, paying two points for the new loan. The points are financed with the new loan.

a. What is the loan balance of the original mortgage after five years?
b. What is the total amount of the new loan?
c. What is the APR of the new loan?
d. Should you proceed with the refinance?

3-10. Kellerman Farm Supply, Inc., has both cash sales and credit sales. Half of each month's sales is for cash and all credit sales are collected one month later. Sales for April, May, June, and July are $35,000, $44,000, $40,000, and $30,000, respectively. Cash payments for May, June, and July are $45,000, $32,000, and $28,000. May's beginning cash balance is $2,000, which is the desired minimum. Ignoring interest expense, complete the cash budget for May, June, and July. If these are Kellerman's most difficult cash-flow months during the year, how much would you recommend it request from its bank for a line of credit?

4
Capital Budgeting and Leasing

The subjects of capital budgeting and leasing both deal with the allocation of funds to obtain the use of capital assets. Both subjects involve financial decision making and can impact the profitability, liquidity, and solvency of the firm. **Capital budgeting** is defined as the process of planning expenditures on assets whose returns will extend beyond one year (i.e., expenditures on capital assets). The concept of the time value of money as discussed in chapter 2 is essential to the more sophisticated capital budgeting techniques.

Leasing allows the agribusiness manager to use an asset without the burden or the benefits of ownership and usually with a smaller up-front cash outflow than purchasing. A **lease** is a contractual agreement between a lessee (the renter or tenant) and a lessor (the owner or landlord). During the term of the lease, the lessee pays rent to the lessor for the possession and use of an asset. Upon completion of the term of the lease, the rights of possession and use revert to the lessor.

We begin this chapter with a discussion of the important issue of the cost of capital. We will then cover the capital budgeting techniques of the payback method, simple rate of return, net present value, benefit-cost ratio, and internal rate of return are then covered. The chapter ends with a section on the subject of leasing.

■ The Cost of Capital

Capital budgeting decisions often require consideration of a firm's **cost of capital**, defined as the cost of financing a business expressed as an interest rate. The cost of capital is equal to the weighted average of the cost of debt capital and the cost of equity capital. In equation form, this is stated as:

$$K_c = w_d K_d + w_e K_e \tag{4.1}$$

in which K_c is the cost of capital expressed as a percentage,
w_d is the proportion of assets financed by debt,
K_d is the cost of debt expressed as a percentage,
w_e is the proportion of assets financed by equity,
K_e is the cost of equity expressed as a percentage.

The proportions of assets financed by debt and equity (i.e., w_d and w_e) are easily found. All one needs is the firm's balance sheet. Divide total liabilities by total assets to calculate w_d and then divide total owner's equity by total assets to calculate w_e. For example, assume that an agribusiness has total assets of $2,000,000, total liabilities of $800,000, and total owner's equity of $1,200,000, then its w_d is equal to 0.4 and its w_e is 0.6. Obviously, w_d plus w_e will always equal 1.0 because total liabilities plus total owner's equity are equal to total assets.

The cost of debt (K_d) is calculated from the year's beginning and ending balance sheets and income statement. The cost of debt in dollar form is simply the interest expense that the firm pays on its debt. To find the weighted average interest rate paid, divide the total interest expense from the income statement by the average of beginning and ending total liabilities from the balance sheets. Given beginning total liabilities of $600,000, ending total liabilities of $800,000, and total interest expense of $66,500, the cost of debt would be equal to 0.095 or 9.5%:

$$K_d = \frac{\$66,500}{\dfrac{\$600,000 + \$800,000}{2}} = 0.095$$

The cost of equity financing (K_e) is the cost to the business of keeping its owners sufficiently satisfied to remain as owners. It is the minimum rate of return on equity that owners must receive from the business in the form of ordinary income and capital gains.

Unfortunately, estimating the cost of equity for non–publicly traded agribusinesses is difficult. This estimation involves the theoretical concept of an opportunity cost. We assume that owners who behave rationally will retain their equity in a business if they make a long-run return that is equal to or greater than what they could make in their next-best alternative investment. Although it is difficult to pinpoint exactly, each equity holder will have ideas about the value of this opportunity cost, and each business manager should have ideas about what that rate is for the equity holders in his or her business.

Another consideration is that the cost of equity to a business should be higher than its cost of debt. This is true because equity holders take on more risk than debt holders and therefore expect a higher return. Debt holders get reimbursed before equity holders upon liquidation of a business. Consequently, equity holders take on a greater risk of not getting their capital back.

We are now ready to apply the cost-of-capital formula (4.1) to an example. Continuing with the numbers generated so far ($w_d = 0.4$; $K_d = 0.095$; $w_e = 0.6$), all we need is K_e. Assuming that the business owners require a minimum return of 12% on their equity, we can calculate the cost of capital to be 0.11 or 11%:

$$K_c = (0.4)(0.095) + (0.6)(0.12) = 11\%$$

We will return to the cost-of-capital concept when we discuss two capital budgeting techniques, net present value and internal rate of return.

The Payback Method

The payback method of capital budgeting is widely used by business managers. Its primary advantage is that it is easy to apply and understand. The method calculates the number of years that it takes an investment to recoup its initial cash outlay by generating net cash flows. To use the payback method, the investor must know the initial cash outlay required for the investment, and must estimate the future annual net cash flows attributable to the asset(s) being purchased. The investor then begins to sum the net cash flows, constantly comparing the running total to the initial cash outlay. The payback period is the year in which the sum of the net cash flows equals or just exceeds the net cash outlay.

Suppose a farmer is considering buying a new combine that will cost $180,000 cash. The farmer has been relying on custom harvesters in the past and estimates that having his or her own combine will increase his or her net cash flows by $30,000 per year due to timeliness and improved output quality. Adding up $30,000 per year, it takes six years for the combine to recoup its initial cost of $180,000. In other words, the payback period is six years.

Table 4.1. Payback period with uneven net cash flows

Year	Annual net cash flows	Running total
1	$30,000	$30,000
2	30,000	60,000
3	30,000	90,000
4	20,000	110,000
5	20,000	130,000
6	20,000	150,000
7	20,000	170,000
8	20,000	190,000

The net cash inflows do not have to be the same every year to utilize the payback method. Using the combine example again, assume that repairs will increase by $10,000 after the third year. The resulting net cash inflows and their running total are shown in table 4.1.

The payback period is now 8 years because the net cash inflows do not exceed $180,000 until that time. One could argue that the payback period is actually 7.5 years, and this would be true if receipt of the $20,000 in year eight were spread evenly throughout the year. However, the convention in capital budgeting analysis is to assume that cash flows, except for the initial cash outlay, occur at the end of a year instead of the beginning or the middle. This simplifying assumption sometimes parallels reality and sometimes does not.

The investor obviously prefers investments with short payback periods. For example, an investor may require that all of his or her investments have a payback period of five years or less. Setting such a payback requirement is arbitrary and suffers from one of the major disadvantages of the payback method: the method fails to take account of net cash inflows that occur beyond the payback period. An investment with a short payback period may also have a short life, and its financial benefits may end soon after payback is met. Conversely, an investment with a longer payback but with an even longer life has the advantage of providing benefits to the investor further into the future.

Another major disadvantage of the payback method is that it does not consider the time value of money. From chapter 2, the reader should understand that the timing of the inflows can be as important as their magnitude. Another investment that also cost $180,000 but had net cash inflows of $0 in years one through five and of $180,000 in year six would have a payback period of six years, just like the combine above. If only the payback period were considered, then the two investments would look identical and the investor would be indifferent in choosing between them. However, the combine is a better investment since its returns start coming in sooner. Other capital budgeting techniques covered later in this chapter do consider the time value of money.

Simple Rate of Return

The simple rate of return (sometimes referred to as the arithmetic rate of return or return on investment) is equal to the annual net income from an investment as a percentage of the initial outlay needed to acquire the asset(s). Hence, the formula would be:

$$\text{Simple rate of return} = \frac{\text{Net income}}{\text{Initial investment}} \qquad (4.2)$$

Suppose a vegetable packer is considering building a new cooling facility. The total cost would be $1,500,000, and the packer budgets an increase in annual net income of $210,000 resulting from the facility. Applying formula (4.2), the simple rate of return would be 14% ($210,000 ÷ $1,500,000 = 0.14).

Note that net income is used in the simple rate of return method and that net cash inflows are used in the payback method. Net income would include the deduction of depreciation expense, a noncash cost. With the payback method, the objective is to measure how soon the initial cost is recaptured. Depreciation expense is not included since it also deals with recapture; inclusion of depreciation expense in the payback method would be redundant.

The simple rate of return is similar to the payback method in that it also has the disadvantage of not incorporating the time value of money into the analysis. Fortunately, the next three capital budgeting techniques that will be covered do account for the timing of the costs and benefits of an investment.

■ Net Present Value

This capital budgeting technique is widely used by professionals in the industry. Calculation of net present value (NPV) requires a bit more work than calculating the payback period or the simple rate of return, but commonly available computer software programs make finding the NPV much faster and easier than doing it by hand.

Net present value is defined as the present value of an investment's cash inflows minus the present value of its cash outflows. In equation form, this becomes:

$$\text{NPV} = \Sigma\text{PV}_{\text{cash inflows}} - \Sigma\text{PV}_{\text{cash outflows}} \qquad (4.3)$$

The answer is stated as so many dollars of net present value accruing from the investment and can be positive, negative, or equal to zero. One should always be aware of the discount rate that was used to find the present values since changing the discount rate will change the answer; the lower the discount rate the higher the net present value, ceteris paribus. In fact, it is preferable to state the discount rate as part of the net present value; an example would be "a net present value of $20,000 at a discount rate of 8 percent."

The first and most difficult step in determining NPV is to estimate the value and timing of the cash inflows and outflows for the productive life of the asset(s). Within each year, annual outflows should be netted against annual inflows, assuming that they both occur at the end of the year. It is helpful to display these values on a time line with net annual cash outflows expressed as negative numbers and the net annual cash inflows expressed as positive numbers. Usually, the most significant cash outflow will be the initial purchase price, but the NPV technique is versatile enough to accommodate cash outflows at other times as well.

The second step is to calculate the present values of the future cash inflows using present-value-of-a-dollar factors and/or present-value-of-an-annuity factors. Once those inflows have all been discounted to the present, they can be summed together.

The third step is to do the same discounting and summation processes for net cash outflows that may occur in any of the years in the life of the investment. The initial cash outflow, which does not need to be discounted, must be added into the summation of the present values of the cash outflows.

The final step is to use formula (4.3), subtracting the sum of the present value of the outflows from the sum of the present value of the inflows.

A simple example would help to reinforce these steps. Suppose a cattleman has the opportunity to buy a herd of young cows from his neighbor for $40,000. The cattleman estimates he would keep the cows for six more years and then sell them for a salvage value of $20,000. Annual cash revenues attributable to these cows will be $18,000 for each of the six years. Cash expenses (i.e., not including depreciation) will be $10,000 per year. Therefore, net annual cash flows are $8,000 annually for six years. At a discount rate of 12%, what is the net present value of this investment?

First, estimation of the cash inflows and outflows has been done for us. These figures can be displayed on a time line as follows:

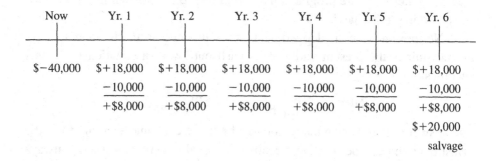

Now	Yr. 1	Yr. 2	Yr. 3	Yr. 4	Yr. 5	Yr. 6
$-40,000	$+18,000	$+18,000	$+18,000	$+18,000	$+18,000	$+18,000
	-10,000	-10,000	-10,000	-10,000	-10,000	-10,000
	+$8,000	+$8,000	+$8,000	+$8,000	+$8,000	+$8,000
						$+20,000
						salvage

To find the present value of the cash inflows, we can treat the annual net inflows of $8,000 as a six-year ordinary annuity. From the appendix, the present value of an annuity factor for six years at 12% is 4.111407. Multiplying this factor by

$8,000, we calculate the present value of the annuity to be $32,891.26. The other cash inflow is the $20,000 salvage value. Multiplying this $20,000 by the present value of a dollar factor for six years at 12%, which is 0.506631, we calculate the present value for this lump sum to be $10,132.62. Adding $32,891.26 and $10,132.62 gives the present value of all cash inflows to be $43,023.88.

The only cash outflow not already accounted for is the initial purchase price, and, since it is already in the present, it will not be discounted. Therefore, the present value of the cash outflows equals $40,000.

Using formula (4.3), we calculate the NPV to be:

$$\text{NPV}_{\text{at } 12\%} = \$43,023.88 - \$40,000 = \$3,023.88$$

Notice that this NPV is positive. This tells the investor that he or she is earning a compound return that exceeds the discount rate by $3,023.88. In other words, the investment pays a compound return of 12% plus the positive NPV. A negative NPV would mean that the investment falls short of earning the discount rate used in the calculation by the number of dollars that the NPV is below zero.

Since changing the discount rate will change the NPV, what rate should an investor use? The answer is that the discount rate should be equal to the investor's cost of capital as defined in formula (4.1). An investor and/or business should consider only investments that will return at least the investor's cost of capital, that is, only investments that have positive NPVs when the cost of capital is used as the discount rate.

NPV analysis can be applied to problems that are more complex than the preceding cow-purchase situation. One common agricultural example concerns permanent plantings (orchards and vineyards). With permanent plantings, land is tied up for a long period of time, and several years of cash outflows occur before any net inflows are realized.

Let's tackle just such a problem involving a peach orchard. Through extensive budgeting, a farmer estimates that a fifty-acre peach orchard will require initial cash outflows of $25,000 to prepare the soil, buy and plant the trees, and install new irrigation equipment. In this example the farmer already owns the land. Therefore, the cost of the land represents a sunk cost, does not entail a cash outflow, and is not included in the analysis. If the farmer were buying land on which to plant the peaches, then the cash cost of the land would be included. This initial cash outflow and the net cash flows for a twenty-year time horizon are as follows:

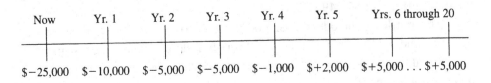

Now	Yr. 1	Yr. 2	Yr. 3	Yr. 4	Yr. 5	Yrs. 6 through 20
$-25,000	$-10,000	$-5,000	$-5,000	$-1,000	$+2,000	$+5,000 ... $+5,000

The cash outflows of $10,000 in year one and $5,000 per year in years two and three represent expenditures to care for the young trees when no harvestable crops are produced. In years four and five, production will occur at one-half and three-quarters of full production, respectively. These peaches will generate cash revenues that will help pay some of the cash expenditures. The net cash flow is projected to be $-$1,000 in year four and $+$2,000 in year five. In years six through twenty, the orchard will be at full production, and the annual net cash inflows are projected to be $5,000. After twenty years of life, the potential of reduced yields and the development of new varieties usually make it necessary to consider the replacement of an orchard. Consequently, we will assume that the farmer does pull out the peach trees after year twenty and that there is no salvage value at that time.

If this farmer has a cost of capital of 9%, what is the NPV of investing in this orchard? The procedure is identical to the simpler cow-purchase problem. The first step, estimating inflows and outflows, has been accomplished and is summarized in the above time line. The second step requires that all cash inflows must be discounted to the present and added together. It is important to discount the inflows first and then sum up all the present values second. It would be very wrong to reverse the order and add up to future values first and discount the sum second.

The first net cash inflow that appears in the time line is the $2,000 that occurs at the end of year five. It is a lump sum, not an annuity, so we must use the present value of a dollar table in the appendix:

$$PV_{at\ 9\%} = (\$2,000)(0.649931) = \$1,299.86$$

The $5,000 inflows in years six through twenty is a fifteen-year annuity, and we should use the present-value-of-an-annuity table for 9% and fifteen years:

$$PV_{at\ 9\%} = (\$5,000)(8.060688) = \$40,303.44$$

However, this $40,303.44 is the value of the annuity at its beginning, the start of the sixth year. The $40,303.44 does not represent the value of the annuity today. Realizing that the start of the sixth year is the same point in time as the end of the fifth year, we must bring this lump sum back from the end of the fifth year to the present using the present value of a dollar. The present-value-of-a-dollar factor at 9% for five years is 0.649931:

$$PV_{at\ 9\%} = (\$40,303.44)(0.649931) = \$26,194.46$$

Adding the inflows of $1,299.86 and $26,194.46 gives a total present value of the cash inflows of $27,494.32.

The outflows include the $25,000 initial cost, which is already in the present. The outflows in years one through four are discounted using the present value of a dollar since they are not the same dollar value in each year and are not an annuity:

$$
\begin{array}{lllll}
PV_{\text{at } 9\%} & = & (\$10,000)(0.917431) & = & 9,174.31 \\
PV_{\text{at } 9\%} & = & (\$5,000)(0.84168) & = & 4,208.40 \\
PV_{\text{at } 9\%} & = & (\$5,000)(0.772183) & = & 3,860.92 \\
PV_{\text{at } 9\%} & = & (\$1,000)(0.708425) & = & \underline{708.42} \\
& & \text{Total PV} & = & \$17,952.05
\end{array}
$$

The total present value of cash outflows is $25,000 plus $17,952.05 which equals $42,952.05. We now have both parts of the NPV formula (4.3):

$$
NPV_{\text{at } 9\%} = \$27,494.32 - \$42,952.05 = -\$15,457.73
$$

This NPV is a negative number that shows that the peach orchard does not cover the 9% cost of capital and is not an acceptable investment.

Benefit-Cost Ratio

Forming a benefit-cost ratio (B/C) utilizes the same two elements as NPV—the present value of the cash inflows (the benefits) and the present value of the cash outflows (the costs). The benefit-cost ratio is simply the quotient of these two elements instead of the difference as in NPV:

$$
\frac{B}{C} = \frac{\Sigma PV_{\text{cash inflows}}}{\Sigma PV_{\text{cash outflows}}} \tag{4.4}
$$

Instead of going through new examples, the cow-purchase and peach-orchard problems discussed above will be used to illustrate the benefit-cost ratio. The steps would commence with the estimation of cash inflows and outflows, and then follow with the discounting of these cash flows, the summation of all discounted cash inflows, and the summation of all discounted cash outflows. For the cow-purchase problem, the ratio of the discounted cash inflows to the discounted cash outflows would be:

$$
B/C_{\text{at } 12\%} = \$43,023.88 / \$40,000 = 1.076{:}1
$$

For the peach-orchard problem, the calculations are:

$$
B/C_{\text{at } 9\%} = \$27,494.32 / \$42,952.05 = 0.64{:}1
$$

Generally, the higher the benefit-cost ratio the better. A benefit-cost ratio of greater than one corresponds to a positive NPV meaning the project is more than meeting the cost of capital. Conversely, a B/C less than one, as with the peach orchard, occurs when the NPV is negative, and the investment should be rejected. The unique case in which the B/C is exactly equal to 1:1 corresponds to an NPV of $0. This signifies that the investment has a compound rate of return exactly equal to the firm's cost of capital, which should have been the discount rate that was utilized.

Internal Rate of Return

The internal rate of return (IRR) is similar to the net-present-value and benefit-cost techniques. Again, the first step is to estimate the value and timing of the cash flows over the life of the project. Moreover, the succeeding steps also involve discounting, combining, and comparing the cash inflows and the cash outflows. The distinction is that the objective of internal-rate-of-return analysis is to find the unique discount rate at which the present value of the cash inflows equals the present value of the cash outflows. Stated as a formula:

IRR is the discount rate at which

$$\Sigma PV_{\text{cash inflows}} = \Sigma PV_{\text{cash outflows}} \tag{4.5}$$

Subtracting $PV_{\text{cash outflows}}$ from both sides:
IRR is the discount rate at which

$$\Sigma PV_{\text{cash inflows}} - \Sigma PV_{\text{cash outflows}} = 0$$

Therefore, the internal rate of return is a discount rate at which net present value, the sum of the present value of the cash inflows minus the sum of the present value of the cash outflows, equals zero.

The IRR is the discounted rate of return on an investment, that is, the rate of return considering the time value of money. Unlike the simple rate of return, that looks at just one year, the IRR considers all inflows and all outflows over the life of the project and states the relationship of those inflows to those outflows as a single rate.

The major disadvantage with using the IRR is that its calculation requires an iterative, trial-and-error process that can be tedious. Fortunately, computer software has been developed that dramatically reduces the calculation time.

However, calculation by hand reinforces one's understanding of the nature of the IRR, so we will go over a couple of simple examples. Suppose an agricultural accountant is considering the purchase of a new computer system for his or her office. The initial hardware and software costs would be $16,000. This cost and the net annual benefits over a five-year life are displayed on the following time line:

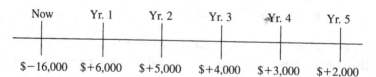

Now	Yr. 1	Yr. 2	Yr. 3	Yr. 4	Yr. 5
$-16,000	$+6,000	$+5,000	$+4,000	$+3,000	$+2,000

First, a beginning discount rate must be selected. Any rate will do. Let's try 12%. The present value of the cash outflows is $16,000. Using the present value of a dollar table in the appendix, the present value of the cash inflows at 12% is as follows:

$$PV_{\text{at } 12\%} = (\$6,000)(0.892857) = \$\ 5,357.14$$
$$PV_{\text{at } 12\%} = (\$5,000)(0.797194) = 3,985.97$$
$$PV_{\text{at } 12\%} = (\$4,000)(0.71178) = 2,847.12$$
$$PV_{\text{at } 12\%} = (\$3,000)(0.635518) = 1,906.55$$
$$PV_{\text{at } 12\%} = (\$2,000)(0.567427) = \underline{1,134.85}$$
$$PV_{\text{cash inflows at } 12\%} = \$15,231.63$$

The NPV at 12% is equal to $15,231.63 − $16,000 = −$768.37. Since this is not equal to zero, the IRR has not been found and is not 12%. A lower rate must be tried since too much interest has been taken out of the cash inflows at 12%. A lower rate would result in a higher present value for the cash inflows. Let's try 9%:

$$PV_{\text{at } 9\%} = (\$6,000)(0.917431) = \$\ 5,504.59$$
$$PV_{\text{at } 9\%} = (\$5,000)(0.84168) = 4,208.40$$
$$PV_{\text{at } 9\%} = (\$4,000)(0.772183) = 3,088.73$$
$$PV_{\text{at } 9\%} = (\$3,000)(0.708425) = 2,125.28$$
$$PV_{\text{at } 9\%} = (\$2,000)(0.649931) = \underline{1,299.86}$$
$$PV_{\text{cash inflows at } 9\%} = \$16,226.86$$

The NPV at 9% is equal to $16,226.86 − $16,000 = +$226.86. A positive NPV such as this means that not enough interest has been taken out of the cash inflows. The IRR is therefore somewhere between 9% and 12%.

Trying 10% as the discount rate yields an NPV of −$117.08 (calculations are not shown). Since the NPV is negative at 10%, the IRR must be less than 10%. Therefore, the trial-and-error method has narrowed the IRR range to between 9% and 10 %. We might next use 9.5% as the discount rate. Unfortunately, the compound interest tables in the appendix are valid only for whole-number interest rates. To proceed, we must develop our own present-value-of-a-dollar factors using:

$$\text{Present value of a dollar factor} = \frac{1}{(1 + i)^n}$$

Substituting 9.5% for i and one through five for n in this formula gives the factors found in the following:

$$PV_{\text{at } 9.5\%} = (\$6,000)(0.913242) = \$\ 5,479.45$$
$$PV_{\text{at } 9.5\%} = (\$5,000)(0.834011) = 4,170.06$$
$$PV_{\text{at } 9.5\%} = (\$4,000)(0.761654) = 3,046.62$$
$$PV_{\text{at } 9.5\%} = (\$3,000)(0.695574) = 2,086.72$$
$$PV_{\text{at } 9.5\%} = (\$2,000)(0.635228) = \underline{1,270.46}$$
$$PV_{\text{cash inflows at } 9.5\%} = \$16,053.31$$

The NPV at 9.5% is +$53.31, so a slightly higher rate must be tried. Perhaps 9.6%? One can see that this becomes a tedious process, and any time spent on increased accuracy is of questionable worth. The actual IRR is 9.655427%, which was found using a computer program and is verified to be the IRR in the following:

$$PV_{\text{at } 9.655427\%} = (\$6,000)(0.911948) = \$\,5,471.69$$
$$PV_{\text{at } 9.655427\%} = (\$5,000)(0.831648) = 4,158.24$$
$$PV_{\text{at } 9.655427\%} = (\$4,000)(0.75842) = 3,033.68$$
$$PV_{\text{at } 9.655427\%} = (\$3,000)(0.691639) = 2,074.92$$
$$PV_{\text{at } 9.655427\%} = (\$2,000)(0.630739) = \underline{1,261.48}$$
$$PV_{\text{cash inflows at } 9.655427\%} = \$16,000.01$$

$$NPV_{\text{at } 9.655427\%} = \$16,000.01 - \$16,000.00 = \$0.01$$

This NPV is close enough to zero for us to be confident that the IRR has been found. Any NPV less than five or ten dollars should be reasonably accurate. The student should always state that the IRR has been found and what rate it is equal to. In this example, we have already done so, but, for emphasis, let's repeat that the IRR of the computer purchase is equal to 9.655427%.

A second IRR example will illustrate the procedure when the cash flows include annuities. We will use the same computer purchase example, but will assume that the cash inflows are the same each year as shown in this time line:

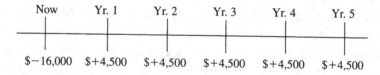

Again, a beginning discount rate for the trial-and-error process must be set. Starting at 10% and using the present value of an annuity table in the appendix, we calculate the present value of the cash inflows:

$$PV_{\text{of cash inflows at } 10\%} = (\$4,500)(3.790787) = \$17,058.54$$

From formula (4.3), the NPV at 10% is +$1,058.54 (i.e., $17,058.54 present value of the cash inflows minus $16,000 present value of the cash outflows). Since this is positive, the IRR must be at a higher rate since more interest must be removed from the cash inflows. Let's try 13%:

$$PV_{\text{of cash inflows at } 13\%} = (\$4,500)(3.517231) = \$15,827.54$$

$$NPV_{\text{at } 13\%} = \$15,827.54 - \$16,000 = -\$172.56$$

Now the NPV is negative, and too much interest has been taken out of the inflows The IRR must be lower than 13%.

Trying 12%:

$$PV_{\text{of cash inflows at } 12\%} = (\$4,500)(3.604776) = \$16,221.49$$

$$NPV_{\text{at } 12\%} = \$16,221.49 - \$16,000 = +\$221.49$$

The IRR must be between 12% and 13%. To develop present value of an annuity factors for interest rates not listed in the appendix, the following formula is used:

$$PV \text{ of an annuity} = \frac{[1 - (1 + i)^{-n}]}{i} \qquad (4.6)$$

At an i of 0.125 and n of 5, the present-value-of-an-annuity factor is 3.560568. The present value of the cash inflows and the NPV are:

$$PV_{\text{of cash inflows at } 12.5\%} = (\$4,500)(3.560568) = \$16,022.56$$

$$NPV_{\text{at } 12.5\%} = \$16,022.56 - \$16,000 = \$22.56$$

Since $22.56 is relatively close to zero, the trial-and-error process is narrowing in on the correct IRR. Again, we resort to the computer program to shorten the process, although further iterations certainly could be performed manually. The correct IRR is 12.55733%, and using formula (4.6) shows this to be true:

$$PV_{\text{of cash inflows at } 12.55733\%} = (\$4,500)(3.555556) = \$16,000.00$$

$$NPV_{\text{at } 12.55733\%} = \$16,000 - \$16,000 = \$0.00$$

Again for emphasis: the IRR has been found and it is equal to 12.55733%.

An investor prefers investments with high internal rates of return. After IRR has been calculated, it should be compared to the investor's and/or business's cost of capital. If the IRR is above the cost of capital, and cash flows behave as they have been projected, then the investment will increase the investor's (business's) profits in the future.

■ Income Taxes and Capital Budgeting

The consequences of income taxes should be taken into account in capital budgeting analysis since the payment of income taxes constitutes a cash outflow. The most efficient procedure is to measure the after-tax benefits and costs of an investment by multiplying the taxable cash revenues and cash expenses by one minus the tax rate, denoted by t:

(Taxable cash revenues)$(1 - t)$ = after-tax cash revenue

(Taxable cash expenses)$(1 - t)$ = after-tax cash expense or

(Taxable net cash income)$(1 - t)$ = after-tax net cash income

The analyst should realize that not all cash inflows and outflows of an investment are taxable, and only multiply $(1 - t)$ by those that are.

From a tax point of view, depreciation expense is a special case. It is not a cash outflow, and, as such, has not been included in any of our capital budgeting procedures except for the simple rate of return. But, depreciation is a taxable expense, so it affects income taxes. The higher the depreciation expense, the more the investor can deduct as a taxable expense, the lower the taxable income, and the lower the taxes paid. Depreciation expense shields income from taxation and reduces the cash outflows of the tax-paying investor. Therefore, tax savings due to depreciation expense should be included in payback method, NPV, B/C, and IRR analyses even though depreciation expense itself is not included as an outflow. Benefits due to the depreciation tax shield are calculated by multiplying the depreciation expense by the tax rate. If the tax rate is 30%, then every dollar of depreciation reduces taxable income by a dollar and reduces taxes by thirty cents.

Consider once again the computer-purchase example in which $16,000 is the initial cash outlay and the net cash inflows amount to $4,500 per year for five years. The internal rate of return on this investment was calculated to be 12.557833%. Now let's find the IRR with the consequences of income taxes included. The $16,000 cash outlay does not need to be adjusted for taxes because it is not a revenue or an expense at the time of purchase. Assuming that the $4,500 per year is the difference between taxable cash revenues and taxable cash expenses, the $4,500 should be modified for tax implications in capital budgeting analysis. Given a tax rate for the firm of 30% once again, the after-tax cash flow is equal to ($4,500)(1 − .30), which is $3,150. Next the tax savings resulting from depreciation expense must be calculated. We will use the Modified Accelerated Cost Recovery System (MACRS) of the Internal Revenue Service. The annual depreciation expense on the computer is equal to the purchase price times a percentage prescribed by the IRS. These percentages that apply to the computer and the resulting depreciation expense are shown in columns 3 and 4 of table 4.2.

Table 4.2 also shows the tax savings from depreciation (column 5), which is equal to the annual depreciation expense multiplied by the 30% tax rate. This savings is then added to the after-tax cash flow ($3,150 in column 6) to find the total after-tax cash flow (column 7). The internal rate of return can be found using the net inflows in column 7 and the initial cash outflow of $16,000. A computer spreadsheet program calculates the after tax IRR of this investment to be 8.83375%. Notice that this is less than the 12.55733% IRR obtained before income taxes had been considered. The after-tax IRR is lower because the burden of the payment of taxes outweighs the tax savings derived from the depreciation tax shield.

Table 4.2. Tax implications in capital budgeting: a computer purchase example

Year	Purchase price	MACRS percent	Depreciation expense	Tax savings from depreciation	After tax cash flow (excl. depr.)	Total after-tax cash flow
(1)	(2)	(3)	(4)	(5)	(6)	(7)
1	$16,000	15.00	$2,400	$720	$3,150	$3,870
2	16,000	25.50	4,080	1,224	3,150	4,374
3	16,000	17.85	2,856	857	3,150	4,007
4	16,000	16.66	2,666	800	3,150	3,950
5	16,000	16.66	2,666	800	3,150	3,950
6	16,000	8.33	1,333	400		400

■ Other Considerations Regarding Capital Budgeting

Now that the techniques of capital budgeting have been learned, a few words should be added about their application. Considering one project at a time and checking whether that project meets some previously determined standard or hurdle is called a **screening decision**. Screening decisions would include checking to see if an investment has a positive NPV using the cost of capital as the discount rate, if the IRR is greater than the cost of capital, or if the payback period is less than a required number of years.

Although examining one investment at a time is of value, greater benefits accrue when a set of alternative investments is analyzed, compared, and ranked. This analysis forms the basis for a **preference decision**. At any point in time, an agribusiness manager has a large number of potential projects that could be undertaken. The projects that are funded will impact the performance of the firm for years into the future. If capital budgeting decisions are going to be anywhere near optimal, then the implementation of the highest ranking investments is imperative.

A problem arises when using net present value for ranking investments of different size. A combine costing $180,000 and a computer system costing $16,000 are clearly not the same size. Both have positive net present values at the firm's cost of capital, and the combine by virtue of its size has the larger NPV. Consequently, the combine would be chosen if dollars of net present value were the decision criterion. However, the benefit-cost ratio and the internal rate of return may be greater for the computer system than for the combine. B/C and IRR are better techniques to use for ranking and preference decisions in this case, but this statement is subject to two very important conditions. These two conditions are that other investments must exist for the $164,000 that remains after the computer has been funded, and that these other investments must have higher B/Cs and IRRs than the combine.

Differences in years of life between investments creates another problem with ranking. Assume two investments—one with a five-year life and one with a twenty-year life. The five-year investment has the higher IRR and is selected. But what happens in years six through twenty? A common solution to this problem is to assume that the five-year investment and its IRR can be repeated four times. In situations where this assumption is not valid, it is possible to compare the investments by calculating their NPVs and converting those NPVs into annuities. Conversion to annuities is accomplished by multiplying the NPV times the amortization factor corresponding to the cost of capital and the investment's years of life. The investment with the higher annuitized value should be chosen.

As previously mentioned, the most difficult task in capital budgeting is estimating the value and timing of the cash flows. The analyst must conduct research to gather as much knowledge as possible about the engineering, productivity, cost savings, increased revenues, resource requirements, useful life, upkeep, and other factors of the proposed projects. No one can exactly predict the future, but if the analyst is objective and impartial in his or her budgeting, then the best investment(s) should emerge. In small agribusiness firms, the analyst is often the same person who proposes the projects and makes the final decisions. That person should guard against bias. In large firms, bias is often minimized by finance department personnel completing the capital budgeting analyses on projects that are proposed by personnel from the other departments within the firm. Upper management then makes the decisions about which projects are undertaken.

In conclusion, capital budgeting is a powerful tool of the financial manager. The objective of the analysis is to determine the long-run *profitability* of investments. To this end, cash purchases are assumed. But many investments are financed with borrowed funds. If an investment is found to be profitable using capital budgeting techniques, the financial manager should then determine the *financial feasibility* of the investment. Financial feasibility refers to whether or not the firm can afford the down payment and the periodic payments on loans associated with the investment. This issue will be discussed in chapter 6 under the topic of repayment capacity.

■ Leases and Leasing

Agribusinesses commonly lease both real and personal property. Leasing allows a firm to use an asset to enhance performance usually with a smaller up-front commitment of funds than purchasing the asset. Leasing, however, is not the better alternative in all situations. There are several variables that affect whether an asset should be purchased or leased. These variables include whether the purchase will be cash or credit, the interest rate and down payment required for a credit purchase, the depreciation on the asset (if depreciable), the terms of the lease contract, the firm's liquidity position, its cost of capital, and its tax rate.

Real Property

Over half the farmers in the United States lease all or some of their land, buildings, and facilities. Also, agribusinesses in the input-supply and output-marketing sectors frequently lease real property. Leasing real property is a costly decision, so the business manager should give that decision thorough deliberation. But leasing is not as costly or as momentous a step as buying real estate. Leasing offers an alternative to firms that cannot afford to purchase or prefer to lease because it increases the flexibility of their operations.

Like any contract, leases are negotiated. The lessor (landlord) is usually responsible for drawing up a written lease, often paying an attorney to complete the task. But the lessee (tenant) should not sign a lease until the lessee has read it and is sure that all provisions are acceptable. What provisions are found in most land leases? Certainly, the term of the lease will be specified. Common terms are one, three, and five years, but many other terms are possible—there are even ninety-nine-year leases. A month-to-month lease is one in which either party can terminate the lease by giving the other party one month written notice. Two other essential provisions are the rent (or some method of calculating the rent) and the timing of rental payments. The property is described in the lease, and the responsibilities of each party (i.e., who pays for what and who performs which functions) are itemized. It is imperative that the listing of responsibilities is as exhaustive as possible to avoid misunderstandings. The lease will also include sections about what happens in the case of default by the tenant and whether or not subletting is allowed.

Both cash leases and share leases are found in farming. A **cash lease** is one in which cash rent is paid, with all or part of the rent paid in advance. In a **share lease**, the landlord receives a portion (25%, 33%, 50%, etc.) of the output at harvesttime as payment. Sometimes the landlord's portion is sold right along with the tenant's portion, and sometimes the landlord sells the landlord's portion independently. Share leases are used for both crop and livestock enterprises, but crop share leases are more common. The portion of the output that the landlord receives varies by area and by enterprise. Share leases are often used for grain, oilseed, hay, and cotton farms, but cash leases predominate in vegetable, fruit, and nut farms. With a share lease, risks are shared by both the landlord and the tenant. With a cash lease, the tenant takes on all production and price risk.

Personal Property

Equipment and breeding stock are the two main types of personal property leased in agriculture, with equipment leases being more prevalent. Over the last couple of decades, personal property leases have become more readily available in the marketplace.

A distinction is made between operating leases and capital leases. An **operating lease** is a short-term contract in which the lessee agrees to pay rent to the lessor based on the amount of time the asset is used. Examples are renting equipment by the hour, the day, the week, or the month. In livestock leases, a bull might be rented for a breeding season. Operating leases are worthwhile when the asset is specialized in nature and only needed for a short period of time. A problem with timeliness can arise if several agribusinesses in an area need to rent the specialized asset at a crucial time, and the lessor only has a small number of the assets available.

The key difference between an operating lease and a capital lease is the length of the term. A **capital lease** is a noncancelable, long-term (i.e., greater than one year) contract in which the lessee pays rent on a monthly or annual basis and maintains possession of the asset even when it isn't in use. Capital leases on cars and trucks are frequently advertised on television. But capital leases are used for an even wider variety of assets including computers, copiers, tractors, forklifts, dairy cows, and horses. The lessee can arrange for such a lease through a dealer, a commercial bank, a Farm Credit System office, or a leasing company. A leasing company is in the business of purchasing and then leasing out all types of assets to satisfy the specific needs of lessees.

The issue of whether a contract is a capital lease or a purchase contract impacts both the lessee's financial statements and tax returns. To distinguish between capital leases and purchases there are separate sets of rules drawn up by the Financial Accounting Standards Board for financial statements and the tax code for tax returns. The Internal Revenue Service says that an agreement is a sales contract and not a lease if any of the following is true:

1. The agreement applies part of each payment toward an equity interest you will receive.
2. You receive title to the property after you pay a stated amount of required payments.
3. You must pay, over a short period, an amount that represents a large part of the price you would pay to buy the property.
4. You pay much more than the current fair rental value of the property.
5. You have an option to buy the property at a small price compared to the value of the property at the time you can exercise the option. Determine this value at the time of entering into the original agreement.
6. You have an option to buy the property at a small price compared to the total you must pay under the lease.
7. The lease designates part of the payments as interest, or part of the payments are easy to recognize as interest.[1]

In the example that follows dealing with whether it is better to acquire personal property through purchase or through a capital lease, we have assumed that the capital lease qualifies as a lease under the above set of rules.

Suppose a feed supply store is considering the acquisition of a new delivery truck. The manager of the store has analyzed the profitability of the truck using capital budgeting techniques and is now trying to decide between a credit purchase and a capital lease. To make this decision, we will focus on only those cash outflows and inflows that differ between the two alternatives, for only those are relevant to the decision. The objective is to find the alternative that exhibits the lowest present value of the net cash outflows (i.e., the least "cost" in today's dollars).

The $30,000 truck can be leased for five years at an annual rental payment of $7,000. Rent is paid at the time the lease is negotiated and then at the beginnings of the following four years. Rent is a tax-deductible expense, and the feed store is in the 35% tax bracket. For every dollar spent on rent, the feed store reduces its taxable income by a dollar, and reduces its taxes by thirty-five cents. The tax benefit is assumed to accrue at the ends of the years. A cost of capital of 12% is used to calculate the present value of the net cash flows of the capital lease in table 4.3.

In table 4.3, the net cash flow (column 4) is equal to the rent (column 2) added to the tax benefit. The present value in dollars (column 6) is equal to the net cash flow multiplied times the present value of a dollar factor for 12% (column 5). The $-19,431$ at the bottom of column 6 represents the present value of only the relevant costs and benefits. Costs, such as fuel, labor, repairs, and license fees, as well as the benefits from the truck, are assumed to be the same to the feed store whether the truck is rented or purchased.

The truck could also be purchased using a five-year loan with amortized payments at 10% interest. A down payment of 30% ($9,000) is required. Under this alternative, the truck is a depreciable asset belonging to the feed store. Remember that depreciation expense is not a cash outflow, but it does reduce the owner's tax obligation. Conversely, the principal portion of the loan

Table 4.3. Capital lease of delivery truck: Present value of relevant net cash flows

Year	Rent	Tax benefit	Net cash flow	Present value factor	Present value
(1)	(2)	(3)	(4)	(5)	(6)
0	$-7,000		$-7,000	1.0	$-7,000
1	-7,000	$+2,450	-4,550	0.892857	-4,063
2	-7,000	+2,450	-4,550	0.797194	-3,627
3	-7,000	+2,450	-4,550	0.711780	-3,239
4	-7,000	+2,450	-4,550	0.635518	-2,892
5		+2,450	+2,450	0.567427	+1,390
			Present value of cash flows		$-19,431

payment is a cash outflow but is not a tax-deductible expense, so it has no tax consequences. The interest portion of the loan payments is both a cash outflow and a tax-deductible expense. Table 4.4 summarizes the credit-purchase alternative.

In table 4.4, the principal and interest portions of the payments (columns 2 and 3) were found using an equal-payment-plan amortization table as discussed in chapter 2. Each annual payment, including both principal and interest, is equal to $5,540. The depreciation expense (column 4) is based on the IRS's Modified Accelerated Cost Recovery System (MACRS) for five-year property using the 150% declining balance method and the half-year convention. Depreciation expense spills into year six because MACRS is used. The tax benefit (column 5) is the sum of the tax-deductible expenses of interest and depreciation (columns 3 and 4) multiplied by the 35% tax rate. The net cash flow (column 6) is the combination of the down payment and principal (column 2), the interest expense (column 3), and the tax benefit (column 5). As in table 4.3, the cost of capital of 12% was used to convert the net cash flows into present values. The present value of all the relevant cash flows for the credit purchase alternative adds up to be $-19,745.

We can now begin to compare the alternatives of leasing and purchasing on credit. The present value for leasing of $-19,431 is a smaller net outflow than the $-19,745 for the credit purchase, so the leasing alternative looks to be slightly advantageous. However, one important element has been left out of tables 4.3 and 4.4: *It has not been stated what the feed store manager intends to do with the delivery truck at the end of the five-year period.* Under the leasing alternative, the manager can return the truck to the lessor and do without, purchase the truck that has been leased for five years, or lease (or purchase) another truck. Under the credit purchase alternative, the manager can continue

Table 4.4. Credit purchase of delivery truck: Present value of relevant net cash flows

Year	Principal	Interest	Depreciation	Tax benefit	Net cash flow	Present value factor	Present value
(1)	(2)	(3)	(4)	(5)	(6)	(7)	(8)
0	$-9,000				$-9,000	1.0	$-9,000
1	-3,440	$-2,100	$-4,500	$+2,310	-3,230	0.892857	-2,884
2	-3,784	-1,756	-7,650	+3,292	-2,248	0.797194	-1,792
3	-4,162	-1,378	-5,355	+2,356	-3,183	0.711780	-2,266
4	-4,578	-961	-4,998	+2,086	-3,454	0.635518	-2,195
5	-5,036	-504	-4,998	+1,926	-3,614	0.567427	-2,051
6			-2,499	+875	+875	0.506631	+443
					Present value of cash flows		$-19,745

to utilize the truck that is now owned free and clear, can sell that truck and do without, or can sell that truck and purchase (or lease) another truck. Any of these scenarios could be worked into the analysis.

To put the feed store in the same position with regards to the truck at the end of five years, we will assume that the purchased truck is sold for $5,000 cash at that time. This assumption eliminates the sixth year's depreciation, and results in a taxable gain on the sale of the truck. That taxable gain is equal to $2,501, which is the $5,000 sales price minus the book value of $2,499. Like depreciation, taxable gains affect taxes but are not cash flows. These changes are summarized in table 4.5.

Comparing a present value of $−17,848 in relevant net cash outflows for the purchase (with resale) alternative to the $−19,431 for the leasing alternative shifts the decision in favor of a credit purchase. Please realize that this comparison is only valid if the feed store manager actually intends to sell the truck as is presented in table 4.5. The conclusion could be different if the manager has other intentions as well as if there were changes in the rental payment, term of the lease, tax bracket, cost of capital, down payment, loan term, loan interest rate, or depreciation schedule. The beauty of this type of analysis is its flexibility; any changes in the variables could be incorporated to better reflect reality and better aid in decision making.

The decision of whether to lease or buy is often made in favor of leasing because the decision maker does not have the funds for an outright cash purchase or even for a down payment. This is an issue of financial feasibility, that is, can the investor afford to purchase the asset? In the preceding analysis, it was assumed that the $9,000 down payment was available to the feed store manager. If the $9,000 is not available, then the only financially feasible alternative might be to lease the truck even though we show that the purchase option shows a lower present value of the relevant cash flows under the assumptions of table 4.5.

Another factor that can have psychological advantages for leasing versus a credit purchase is the fact that leases are often negotiated with monthly, not annual, payments. A small monthly payment on a lease, even though it must be paid twelve times a year, simply sounds better to many individuals than a larger annual loan payment. This argument is a bit spurious, however, since the loan could easily be set up with monthly payments as well. If monthly payments are made under either option, the manager should still conduct an analysis as outlined in tables 4.3, 4.4, and 4.5. The timing of the cash flows could be set up on a monthly basis and then brought back to the present using a monthly discount rate. Such monthly tables are simple to construct using a spreadsheet computer program, but they are quite large and will not be displayed in this text due to that fact.

Table 4.5. Credit purchase (with resale) of delivery truck: Present value of relevant net cash flows

(1) Year	(2) Principal	(3) Sales price	(4) Interest portion	(5) Depreciation	(6) Taxable gain	(7) Tax benefit	(8) Net cash flow	(9) Present value factor	(10) Present value
0	$-9,000						$-9,000	1.0	$-9,000
1	-3,440		$-2,100	$-4,500		$+2,310	-3,230	0.892857	-2,884
2	-3,784		-1,756	-7,650		+3,292	-2,248	0.797194	-1,792
3	-4,162		-1,278	-5,355		+2,356	-3,183	0.711780	-2,266
4	-4,578		-961	-4,998		+2,086	-3,454	0.635518	-2,195
5	-5,036	$+5,000	-504	-4,998	$+2,501	+1,050	+510	0.567427	+289
						Present value of cash flows			$-17,848

■ Questions and Problems

4-1. A cattle rancher had average total assets of five million dollars and average total liabilities of two million dollars during the preceding year. Her cost of debt for that year was 7.5% and she estimates her cost of equity to be 10%. What was this rancher's weighted average cost of capital expressed as a percentage?

4-2. Calculate the payback period for a dry-bean harvester that requires an initial cash outlay of $250,000. The after-tax net cash flows from this harvester will be $60,000 during the first year; $50,000 for each of the second, third, and fourth years; and $30,000 for the fifth, sixth, seventh, and eighth years. In year eight, the harvester can be sold for an after-tax salvage value of $40,000.

4-3. What is the simple rate of return on an almond-hulling business you can buy for $1,500,000 that will generate a net income of $200,000 per year?

4-4. A farm building costs $54,000 to build today and will earn after-tax net cash flows of $16,000 per year for five years. There is no salvage value on this building at the end of the five-year life, and the farmer's cost of capital is 9%.

  a. What is its net present value?
  b. What is the benefit-cost ratio of this building at a cost of capital of 9%?
  c. Calculate the internal rate of return on this building. Use the trial-and-error method and show all your iterative steps.

4-5. What is the internal rate of return (to the nearest one-half percent) on an investment costing $500,000 and having expected future after-tax net cash flows of:

Year	Net Cash Flow ($)
1	100,000
2	150,000
3	150,000
4	300,000 (includes salvage)

Use the trial-and-error method and write out all your work. Hint: start at 10%.

4-6. Your father has inherited lots of money, and he has always wanted to own a farm. He is considering the purchase of several farm properties. Additionally, your father expects to sell the farm and retire after owning the farm for a full fifteen years.

One of the farms he is looking at has a purchase price of $800,000, after-tax development costs of $40,000 per year for the first five years, and after-tax net cash inflows of $35,000 per year for years six through fifteen. To determine the selling price of the farm at the end of the fifteenth year, assume that the farm will increase in value at a rate of 5% per year.

Using a 7% cost of capital, calculate the net present value and benefit-cost ratio of this farm investment. Show your work. Explain whether or not and why you should keep this farm among your investment alternatives.

4-7.   Within a recent tax year, a farmer had taxable cash revenues of $120,000, taxable cash expenses of $70,000, and depreciation expenses of $24,000. The applicable tax rate was 36%. With the tax savings from depreciation included, what was the total after-tax cash flow for the year?

4-8.   An agricultural accounting firm needs three new copy machines. If it leases the copiers on a five-year lease, five rental payments of $21,000 each will be made at the beginnings of the years. The firm is in the 30% tax bracket and has a cost of capital of 9%.

a.   Prepare a table similar to table 4.3 to find the present value of the cash flows for the lease option.

b.   The accounting firm could purchase the copiers on credit. The original cost for all three machines is $82,500, and the total down payment will be $22,500. Amortized loan payments of $15,828 each will be made at the ends of years one through five, and the loan carries an interest rate of 10%. Depreciation expense in years one through six will be: $12,375, $21,038, $14,726, $13,745, $13,745, and $6,872, respectively. The tax rate is once again 30%, and the cost of capital is 9%. Prepare a table similar to table 4.4 to find the present value of the cash flows for the credit purchase option.

c.   Discuss which option the accounting firm should choose— leasing the copiers or buying them on credit. Include any other factors that might influence their decision in your discussion.

■ Note

1. Internal Revenue Service, *Farmer's Tax Guide* (1998), 26.

5
Financial Statements

Well-prepared financial statements provide a wealth of information about a business. That information is of interest to the current owners and managers of the business as well as to potential investors and current or potential lenders. Even when a firm's financial statements are prepared by a professional accountant, the firm's managers should diligently study those financial statements. They should understand every entry on each statement and how the statements coordinate. Financial statements provide information that is essential in making decisions that affect the firm.

The financial statements discussed in this chapter are the income statement, the balance sheet, the statement of owner's equity, and the statement of cash flows, with more emphasis being placed on the first two. It is assumed that the student has completed at least one class in accounting and understands the basic accounting cycle.

■ Generally Accepted Accounting Principles and the Farm Financial Standards Council

Professional accountants prepare financial statements following Generally Accepted Accounting Principles (GAAP). GAAP includes the following basic principles:

1. Business entity—a firm is separate from its owners
2. Going concern—the firm will continue to be in business
3. Matching—revenues should be aligned with the expenses that generate those revenues
4. The stable dollar—accountants do not account for inflation or deflation
5. Historical cost—assets will be recorded at their original cost
6. Full disclosure—all relevant and significant information will appear on the entity's financial statements[1]

These and other principles are interpreted and given substantial authoritative support by the Financial Accounting Standards Board (FASB) through its Statements of Financial Accounting Standards. Nonfarm agribusinesses varying in size from large international food processors to small local chemical supply companies would have their financial statements prepared according to GAAP.

Many farm businesses do not follow GAAP, however. A primary reason is that farmers can (and often do) elect to keep records and file their income taxes using the **cash system of accounting**. Because of this, their financial statements are prepared on the cash system whereas GAAP requires the **accrual system of accounting**. The major difference between the cash and accrual systems deals with the timing of the recognition of revenues and expenses. Under the cash system, revenues are counted in the period of time when they are received, and expenses are counted in the period of time when they are paid. Under the accrual system, revenues are counted in the period earned, that is, when the sale transaction has taken place, and expenses counted in the period incurred, that is, when an obligation to pay has transpired. The accrual system does a much better job than the cash system of matching revenues for time periods with the expenses that it took to generate those revenues. Another common deviation from GAAP for farm and ranch financial statements is in the method of valuation of assets on the balance sheet— growing crops, crop and livestock inventories, raised breeding stock, equipment, and real property. Farm and ranch balance sheets often use the market value of assets while GAAP requires that valuation of assets be based on historical cost.

What had developed in farm accounting during most of this century was a confusing situation in which no single set of standards prevailed. Everyone prepared financial statements using their own set of rules as long as they could get away with what they were doing. The lack of farm accounting standards became especially apparent during the farm financial crisis of the 1980s. In an attempt to ameliorate the problem, the Farm Financial Standards Task Force (later the Farm Financial Standards Council) was established in 1989. This group of volunteer producers, accountants, lenders, academicians, and representatives of farm organizations published its first *Financial Guidelines for Agricultural Producers* in 1991. Later revisions have followed. A primary intention of the *Guidelines* is to "promote uniformity in financial reporting for agricultural producers by presenting methods for financial reporting which are theoretically correct and technically sound."[2]

The Farm Financial Standards Council (FFSC) recommends that agricultural producers use GAAP if possible. Yet, in the many cases where GAAP is not used and will not be used in the foreseeable future, the *Guidelines* serves as a viable substitute. This text will attempt to follow the *Guidelines* in its discussion of financial statements. The reader should realize that the examples will therefore apply to farms and ranches, not to agribusinesses in the input supply and output marketing sectors.

■ The Income Statement

The underlying equation of the income statement is that revenues (sales) minus expenses (costs) equals net income (profits). Revenues, expenses, and net income are flow concepts because they are measured over a period of time. Income statements often pertain to a year's length of time, but monthly, quarterly, and semiannual income statements can be prepared as well.

An income statement measures the *performance* of a business—the performance of assets, management, and labor. If one assumes that the goal of the owners of a business is to maximize profits over the long run, then the income statements over a period of time are the instruments used to judge the attainment of this goal. Net income, the colloquial bottom line, is added to owner's equity at the end of the accounting cycle. Therefore, the goal of long-run profit maximization is consistent with the goal of maximizing the growth in equity.

An income statement for a diversified farm, All American Farms, appears in table 5.1. All American Farms is a sole proprietorship that keeps records under the cash system. The income statement only applies to the farm business. Any personal income and expenses are not included.

Table 5.1. All American Farms' cash system income statement for the year ended December 31, 1998

Soybean sales	$196,000	
Corn sales	183,000	
Calf sales	72,000	
Other farm income	5,000	
Gross revenues		$456,000
Rent	$50,000	
Fertilizer	48,000	
Labor and benefits	45,000	
Seed	37,000	
Chemicals	33,000	
Repairs and maintenance	26,000	
Fuel, oil, and grease	14,000	
Purchased feed	22,000	
Veterinary and medicine	3,000	
Utilities	6,000	
Depreciation expense	52,000	
Property taxes	5,000	
Office expense	3,000	
Operating expenses		344,000
Interest expense	17,000	
Total expenses		361,000
Net farm income from operations		95,000
Gain (loss) on the sale of capital assets	10,000	
Net farm income		105,000
Miscellaneous revenue (expense)	1,000	
Income before income taxes		106,000
Income taxes	14,000	
Net income		$ 92,000

As transactions occur during the year, they are posted to the various income statement accounts found in table 5.1. Other agribusinesses would have other account categories depending on what they produce and the inputs they use. The sums of the transactions within the appropriate accounts are reported on the income statement at the end of the year. At this time, the accounts are emptied out and net income (or loss) is closed to owner's equity on the balance sheet. A positive net income will increase owner's equity and a net loss will reduce owner's equity. In this way, the income statement accounts can be used again in the following year to measure the flows that occur in the business's revenue and expense accounts.

Farm and ranch revenues come primarily from the sale of agricultural commodities such as the soybeans, corn, and cattle for All American Farms. Other possible sources of revenue include government program payments, collection of crop insurance claims, and payment for custom services performed for other farmers. Expenses include a variety of costs of doing business. Depreciation expense, a noncash cost, is counted even under the cash accounting system. Interest paid on business loans is a valid expense, but principal payments are not. Principal repaid is no more counted as an expense than is principal received from a lender counted as a revenue.

In table 5.1 net farm income from operations (NFIFO) is equal to gross revenues minus operating and interest expenses. NFIFO is a better measure of performance than net income because NFIFO measures the results of the normal operations of a business. The words "operations" and "operating" refer to the day-to-day activities of the business in producing and marketing goods and services. NFIFO is next adjusted for gains or losses on the sale of capital assets to calculate net farm income (NFI). Gains are added to NFIFO and losses are subtracted. These gains and losses are equal to the difference between the sales price of a capital asset and its book value. Book value is the historic purchase price of the asset minus the asset's accumulated depreciation up to the time of sale. All American Farms sold capital assets with a book value of twenty thousand dollars for a sales price of thirty thousand dollars and therefore realized a ten-thousand-dollar gain. NFI is next adjusted upwards for miscellaneous revenues and downward for miscellaneous expenses. Miscellaneous revenues and expenses primarily deal with nonfarm investments held by the farm. Finally, income taxes are subtracted to arrive at net income.

Notice that the income statement for All American Farms does not include a cost-of-goods-sold section. This is true for all cash-system and most accrual-system farm and ranch income statements. Agribusinesses in the input supply and output marketing sectors usually report cost of goods sold. From basic accounting, remember that cost of goods sold for a merchandising business is derived using:

> Beginning Inventory
> + Purchases
> _____
> = Goods Available for Sale
> − Ending Inventory
> _____
> = Cost of Goods Sold

For a manufacturing business, which is more similar to a farm than is a merchandising business, the "purchases" that are added to beginning inventory are replaced with the "cost of goods manufactured." Calculating cost of goods manufactured is a complicated procedure involving raw materials inventories, raw materials purchases, direct labor, factory overhead, and work in process inventories. A farm or ranch would have to have a very detailed record keeping system to accurately measure all these elements needed to report cost of goods sold. Also, the listing of expense by category, as in table 5.1, does a better job of reflecting the inputs that were used in production and how much was paid for each input category than would a cost-of-goods-sold section.

Income statements are more useful for analysis purposes if they include quantity and price data. How many tons of soybeans and corn were sold and at what prices? How many head of cattle were sold? How much did the cattle weigh, and what prices were received? How much seed was used, and what price was paid? How was depreciation calculated? These questions can be answered in the actual body of the income statement itself, but it is often more convenient to use notes (also called footnotes) to the income statement. Notes can explain accounting policies, disclose contingencies that may affect the company, or simply describe the raw data behind the summary figures found in the statement.

■ The Statement of Owner's Equity

Don't be fooled by the shortness of the statement of owner's equity (a.k.a. the statement of net worth). It is a very important financial statement because it coordinates the income statement for a year with the beginning and ending balance sheets. The net income (or loss) from the income statement is added to (subtracted from) beginning owner's equity on the statement of owner's equity, and the results of this are then transferred to the balance sheet. If this is not done and if owner withdrawals are not subtracted out, the ending balance sheet will not balance.

As can be observed in table 5.2, the beginning point of the statement of owner's equity is the balance in the owner's equity account at the beginning of the year. The year's net income is then added in and owner withdrawals subtracted out. If the farm had realized a net loss for the year, then the loss would be subtracted from beginning equity. In corporations, withdrawals are called dividends. Withdrawals and/or dividends are not an expense on the income statements as students sometimes think. But they do reduce the equity that is plowed back into the business and they are a cash outflow.

Table 5.2. All American Farms' statement of owner's equity for the year ended December 31, 1998

Owner's equity, beginning of the year	$567,000
Plus: net income	92,000
Less: owner withdrawals	(40,000)
Net additions to retained capital	52,000
Owner's equity, end of the year	$619,000

Owner's equity can also be increased within a given year through the contribution of equity from a source external to the firm. Such external sources could include off-farm income made by the farmer or the farmer's spouse, gifts and inheritances, or funds provided by new partners or stockholders. If any of these funds are invested in the farm, then they will increase the equity on the balance sheet and should appear as an addition on the statement of owner's equity.

■ The Balance Sheet

A firm's balance sheet reports the balances in asset, liability, and owner's equity accounts as of a specific date. Unlike an income statement that measures flows, a balance sheet measures *stock* values (i.e., values at a point in time). The dollar amount of assets must equal the dollar amount of liabilities plus the dollar amount of owner's equity for the balance sheet to balance. Assets are economic resources owned by the firm and can be either tangible or intangible. Liabilities (also called debts) and owner's equity (also called net worth) represent claims against those assets. Liabilities are claims of parties that are external to the firm; owner's equity is the residual claim of the firm's owners, the sole proprietor, the partners, or the stockholders.

Table 5.3 shows two balance sheets for All American Farms. The left-hand column of figures is dated December 31, 1997. It is both the ending balance sheet for 1997 and the beginning balance sheet for 1998 since the end of 1997 is the same point in time as the beginning of 1998. The right-hand column of figures is dated December 31, 1998, and is the ending balance sheet for 1998 (as well as the beginning balance sheet for 1999).

Assets are divided into current and noncurrent categories. **Current assets** (a.k.a. short-term assets) are cash and those assets that will be typically converted into cash within one year. Common current asset categories are cash in both checking and savings accounts, accounts receivable, short-term notes receivable, inventories (harvested crops, market livestock, and supplies), prepaid expenses, and cash in growing crops. **Noncurrent assets** have economic lives of more than one year, meaning they will not typically be sold within a year of the balance sheet date. Common noncurrent asset categories for farms and ranches are breeding livestock (both purchased and raised), machinery and equipment, capital leases, investments in cooperatives, buildings and improvements, and land.

The valuation of assets is an important issue on balance sheets. The two conflicting valuation methods are historical cost and market value. Historical cost means that assets are entered in the books at their original cost and will appear on any subsequent balance sheets at that value. For depreciable assets such as equipment, buildings, and purchased breeding stock, the accumulated depreciation up to the date of the balance sheet will also be shown as a deduction from the original cost. Market value simply refers to the dollar amount that the asset could be sold for in an arms-length transaction between a willing and informed buyer and seller. The significance of the debate between historical cost and market value can be illustrated with an example concerning farmland. All American Farms purchased one hundred acres in 1985 for $150,000. As of December 31, 1997 and 1998, that land is conservatively appraised at $240,000. Should $150,000 or $240,000 be recorded on the balance sheet?

Table 5.3. All American Farms' balance sheet as of December 31, 1997 and December 31, 1998

Assets	(1997)	(1998)
Cash	$ 20,000	$122,000
Crop inventories	50,000	60,000
Accounts receivable	22,000	17,000
Prepaid expenses	6,000	4,000
Cash in growing crops	25,000	25,000
Total current assets	123,000	228,000
Breeding livestock	150,000	150,000
Less accumulated depreciation	(30,000)	(42,000)
Machinery and equipment	380,000	360,000
Less accumulated depreciation	(165,000)	(159,000)
Buildings and improvements	160,000	160,000
Less accumulated depreciation	(66,000)	(72,000)
Land	240,000	240,000
Total noncurrent assets	669,000	637,000
Total farm business assets	$792,000	$865,000

Liabilities	(1997)	(1998)
Accounts payable	$ 32,000	$ 35,000
Short-term notes payable	12,000	26,000
Current portion of term debt	5,000	6,000
Accrued interest	2,000	3,000
Income taxes payable	5,000	5,000
Current portion of deferred taxes	2,000	1,000
Total current liabilities	58,000	76,000
Noncurrent portion of notes payable	25,000	32,000
Noncurrent portion of real estate debt	115,000	111,000
Noncurrent portion of deferred taxes	27,000	27,000
Total noncurrent liabilities	167,000	170,000
Total liabilities	225,000	246,000
Retained capital	504,000	556,000
Valuation equity	63,000	63,000
Total owner's equity	567,000	619,000
Total liabilities and owner's equity	$792,000	$865,000

If drawn up according to GAAP, then the principle of historical cost prevails for land and most other assets. An exception is that GAAP recommends that inventories should be valued at the lower of cost or market value. A problem arises in the industry in that balance sheets with some assets reported at market value are often of greater use to lenders. A lender is interested in what an asset pledged as collateral will bring in the marketplace should that lender have to foreclose. A borrower is often interested in borrowing as much as is possible based on the market value of his or her assets. The Farm Financial Standards Council allows for market value balance sheets to be used, but suggests that historical cost-based data should also be presented. Whichever the system used, the balance sheet should include footnotes that let the reader know.

In table 5.3, inventories for All American Farms were reported at market. Cash in growing crops is the direct costs attributed to the growing crops as of the date of the balance sheet. Breeding livestock includes both raised and purchased animals. Raised breeding stock are valued using a method in which each category of animal is given a base value, and these values are changed infrequently. Purchased breeding stock, machinery, equipment, buildings, and improvements are all reported at cost and the accumulated depreciation on these assets is then subtracted out. Finally, land has been valued at market. One can see that what we call a market-value balance sheet actually uses a combination of valuation methods.

Liabilities are also classified as current and noncurrent. **Current liabilities** (a.k.a. short-term liabilities) will typically be paid within one year of the date of the balance sheet. Accounts payable arise usually from charging goods or services from a vendor with payment being made upon receipt of a bill. A short-term note payable involves a written promise to pay, usually with interest. A line of credit would be an example of a short-term note payable. The current portion of term debt is that part of term-debt principal that will be due within the coming year. Remember that "term debt" refers to loans originally made for a term of more than one year. The current portion of term debt that is shown in current liabilities must be subtracted from the corresponding loans shown in noncurrent liabilities. Accrued interest is an example of an accrued expense that has been incurred but not yet paid, allowing for interest expense to be better matched with its proper time period. Another example of an expense that often must be accrued is salaries. Income taxes-payable represent taxes that have been determined, but have not yet been paid. The current portion of deferred taxes stems from the fact that All American Farms is on the cash accounting system and can defer taxable income by postponing cash receipts or prepaying cash expenses. By deferring taxable income, the farm is also deferring the tax due on that income. Including these deferred taxes as a liability simply discloses that those taxes will be paid within the coming year.

Noncurrent liabilities are due to be paid in more than one year's time. Notes payable have been taken on by All American Farms to finance the purchase of breeding livestock, equipment, and real estate. The noncurrent portion of deferred taxes arises because land has been reported at market value on the asset side of the balance sheet. As has been previously stated, that farmland has increased in market value over its purchase price. When and if that land is sold, a sizable tax will be due on the gain. The noncurrent portion of deferred taxes is an estimate of this contingent tax liability.

The ending owner's equity of $619,000 in table 5.3 comes from the statement of owner's equity (table 5.2). Owner's equity is divided into retained capital and valuation equity. At the business's inception, retained capital is the capital originally contributed by the owners. Since that time, retained capital has been changed by additions of net income, subtractions of net losses, subtractions of owner withdrawals, and additions of new capital contributed by the owners from sources outside the business. Valuation equity measures the after-tax increase in equity over time due to increases in the market value of the farm's assets. The increase in valuation equity does not match dollar for dollar with the appreciation in assets because the noncurrent portion of deferred taxes must be subtracted out. In the case of All American Farms, the valuation equity is calculated by (1) subtracting the original cost of the land ($150,000) from its current market value ($240,000) to calculate the capital gain of $90,000; (2) multiplying the capital gain by a 30% tax rate to find the deferred taxes of $27,000; and (3) subtracting $27,000 in taxes from $90,000 to find a net valuation equity of $63,000.

■ The Accrual-Adjusted Income Statement

A farm or ranch that keeps its books on the cash system can better match revenues and expenses and better measure performance if it prepares an accrual-adjusted income statement. These accrual adjustments are derived by finding the changes in certain account balances between a year's beginning and ending balance sheets. Table 5.4 presents an accrual-adjusted income statement for All American Farms.

Comparing table 5.4 to table 5.1, the cash-system income statement, the first difference that appears is the adjustment for a change in crop inventories. The beginning and ending balance sheets (see table 5.3) show that crop inventories increased from $50,000 to $60,000. This $10,000 increase is added to revenues because it means that the firm is better off inventorywise at the end of the year than at the beginning. Remember that an income statement is trying to measure performance, and this firm performed by increasing its inventory.

Another way to verify that an increase in inventory would act like a revenue involves considering what would happen if a cost-of-goods-sold section

Table 5.4. All American Farms' accrual-adjusted-income statement for the year ended December 31, 1998

Soybean sales	$196,000	
Corn sales	183,000	
+ Increase in crop inventories	10,000	
Calf sales	72,000	
− Decrease in accounts receivable	(5,000)	
Other farm income	5,000	
Gross revenues		$461,000
Rent	$50,000	
Fertilizer	48,000	
Labor and benefits	45,000	
Seed	37,000	
Chemicals	33,000	
Repairs and maintenance	26,000	
Fuel, oil, and grease	14,000	
Purchased feed	22,000	
Veterinary and medicine	3,000	
Utilities	6,000	
Depreciation expense	52,000	
Property taxes	5,000	
Office expense	3,000	
+ Increase in accounts payable	3,000	
+ Decrease in prepaid expenses	2,000	
Operating expenses		349,000
Interest paid	17,000	
+ Increase in accrued interest	1,000	
Interest expense		18,000
Total expenses		367,000
Net farm income from operations		94,000
Gain (loss) on the sale of capital assets		10,000
Net farm income		104,000
Miscellaneous revenue (expense)		1,000
Income before income taxes		105,000
Income taxes paid	14,000	
− Decrease in current portion of deferred taxes	(1,000)	
Income tax expense		13,000
Net income		$ 92,000

were utilized. In the two following hypothetical cost-of-goods-sold sections for a merchandising business, the only difference is that inventory does not change in Case #1 while inventory increases in Case #2:

	Case #1	Case #2
Beginning Inventory	$200,000	$200,000
+ Purchases	700,000	700,000
= Goods Available for Sale	900,000	900,000
− Ending Inventory	200,000	250,000
= Cost of Goods Sold	$700,000	$650,000

Cost of goods sold is a cost (an expense) and is subtracted from revenue to calculate net income. Subtracting out a cost of goods sold of $650,000 (Case

#2) instead of $700,000 (Case #1) would have the effect of increasing net income by $50,000. The effect on net income of the increase in inventory in Case #2 is the same as if there had been a $50,000 increase in revenue. Therefore, on an accrual-adjusted income statement, an increase in an inventory acts like an increase in revenue (a negative expense), and a decrease in inventory would act like an expense (a negative revenue).

Also in the revenue section of table 5.4, the decrease in accounts receivable of $5,000 is subtracted from revenues. On the balance sheets (see table 5.3), accounts receivable changes from $22,000 to $17,000 during 1998. The $22,000 beginning accounts receivable is collected in 1998, and under the cash system it is counted in the gross revenues of $456,000 for 1998 (see table 5.1). Under the accrual system, the $22,000 should have been counted as sales in 1997. Therefore, since we are adjusting in favor of the accrual system, the $22,000 should be subtracted from 1998's revenues. Similarly, the $17,000 ending accounts receivable represents accrual system sales for 1998 and should be added into 1998's revenues. The net effect of subtracting $22,000 and adding $17,000 is a negative $5,000—the adjustment that appears in table 5.4.

The operating expense section of table 5.4 shows two accrual adjustments. An increase in accounts payable of $3,000 is added to operating expenses. What is really happening is that the $32,000 in beginning accounts payable (see table 5.3) is subtracted from operating expenses since it corresponds to expenses incurred in 1997. Likewise, the $35,000 in ending accounts payable is added into operating expenses for 1998. The net effect is the $3,000 increase in expenses for 1998. Prepaid expenses change from $6,000 to $4,000 between the beginning and end of the year. Since this $2,000 of prepaid expenses was not paid in cash during 1998, it is not counted as an expense in the cash-system income statement. However, the $2,000 was used up in 1998 and should be matched with that year under accrual accounting principles.

The increase in accrued interest of $1,000 is added into 1998's interest expense for reasons similar to the increase in accounts payable. Beginning accrued interest of $2,000 is first subtracted out because it belongs to 1997, but the ending accrued interest of $3,000 is added in for 1998.

The last accrual adjustment for All American Farms in 1998 is for a change in the current portion of deferred income taxes. The logic is identical to that for accounts payable and accrued interest.

The accrual adjustments that were found in table 5.4 are not the only ones that can exist. However, all accrual adjustments deal with changes in current-asset accounts (not including cash) and changes in current-liability accounts. Increases in current asset accounts require adjustments that will increase the net income from the firm's cash-system income statement. Conversely, decreases in current-asset accounts reduce net income. Increases in current-liability accounts result in adjustments that decrease net income while decreases in current liabilities will increase net income compared to the cash-system income statement.

The accrual adjusted net income in table 5.4 is $92,000, and that is the same as the net income in table 5.1. This is merely a coincidence—the adjustments, purely by chance, canceled each other out. Accrual adjustments can certainly result in large increases or decreases in net income depending on what has happened to the relevant current asset and current liability accounts during the year. Whatever the result, accrual-adjusted income statements will better report and match revenues and expenses for a year. Lenders often prefer that borrowers provide them with accrual-adjusted income statements.

■ The Statement of Cash Flows

The purpose of the statement of cash flows is to summarize and categorize changes in the cash account that occur within an accounting period. The basic equation behind the statement of cash flows is that ending cash will equal beginning cash plus increases in cash minus decreases in cash. A naive manager whose company has made profits in a year will look at the ending cash account balance and often ask, "well, where did it all go?" This manager is confusing net income with cash. The statement of cash flows answers the question "where did it all go?" as well as the question "where did it all come from?"

Increases and decreases in cash are grouped into three categories on this financial statement: cash flows from operating activities, cash flows from

Table 5.5. All American Farms' statement of cash flows for the year ended December 31, 1998

Cash flows from operating activities		
Cash received from operations	$456,000	
Cash paid for operating expenses	(292,000)	
Cash paid for interest	(17,000)	
Cash paid for income taxes	(14,000)	
Cash received from miscellaneous revenue	1,000	
Net cash provided by operating activities		$134,000
Cash flows from investing activities		
Cash received from sale of equipment	$30,000	
Cash paid for purchase of equipment	(40,000)	
Net cash provided by investing activities		(10,000)
Cash flows from financing activities		
Net increase in short-term debt	$14,000	
Proceeds from term debt	9,000	
Repayment of term debt	(5,000)	
Owner withdrawals	(40,000)	
Net cash provided by financing activities		(22,000)
Net increase (decrease) in cash		102,000
Cash at beginning of year		20,000
Cash at end of year		$122,000

investing activities, and cash flows from financing activities. Operating activities, as stated previously, deal with the day-to-day functioning of the firm, so the cash collection of revenues and the disbursement of cash to cover expenses fit into this section. Investing activities involve cash paid to acquire capital assets and cash received from the disposition of capital assets. Financing activities deal with both debt and equity financing: loan proceeds, loan principal payments, additions to equity other than from net income, and owner withdrawals. The 1998 statement of cash flows for All American Farms appears in table 5.5.

Notice the final three lines in table 5.5. The changes in cash from the three types of activities ($+134,000, $ −10,000, and $ −22,000) combine to a net increase in cash of $102,000. This $102,000, when added to the cash-account balance at the beginning of the year, must equal the cash-account balance at the end of the year. If this is not so, the accountant has made a mistake that may be difficult to find for there are many transactions that influence the cash account within a year's time.

A set of financial statements is often said to be coordinated. The balance sheet is coordinated with the income statements through the statement of owner's equity. The net income for the year is added to the beginning owner's equity and then withdrawals are subtracted. The resulting ending owner's equity is then transferred from the statement of owner's equity to the ending balance sheet to make it balance. The statement of cash flows is also coordinated with the balance sheet through the cash account. The beginning cash plus or minus the changes in cash that are summarized on the statement of cash flows must equal the ending cash balance as found on the balance sheet. Another example of coordination is that the depreciation expense shown on the income statement is added to the accumulated depreciation that appears on the balance sheet.

■ Questions and Problems

5-1. In which financial statement (income statement, statement of owner's equity, balance sheet, or statement of cash flows) would the following items be found? Some items can appear on more than one statement.

a. owner withdrawals
b. accounts payable
c. land
d. depreciation expense
e. accumulated depreciation
f. cash received from the sale of a hay baler
g. wheat sales
h. net income
i. accounts receivable

j.   net increase or decrease in cash
k.   labor expense

5-2. Arrange the items below in their proper order on an income statement
      and calculate net farm income from operations, net farm income, and net
      income.

Operating expenses	$390,000
Interest expense	30,000
Income taxes	40,000
Miscellaneous expense	5,000
Loss on the sale of a capital asset	20,000
Gross revenues	$550,000

5-3. A farm's cash-system net income for a given year is $85,000. The
      beginning and ending balance sheets for the same year show the
      following account balances:

Account	Beginning balance	Ending balance
Accounts receivable	$100,000	$125,000
Inventory	95,000	65,000
Prepaid expenses	9,000	6,000
Accounts payable	20,000	15,000
Accrued expenses	10,000	12,000
Current portion of deferred taxes	28,000	40,000

Calculate the farm's accrual-adjusted net income. Label and show the size
and direction of all adjustments.

5-4. Construct a statement of owner's equity from the information provided:

Beginning owner's equity	$1,350,000
Net income	205,000
Withdrawals	64,000
Equity contributed from spouse's off-farm income	21,000

5-5. A cattle ranch started last year with $64,000 in its cash account. Within
      the year, all the events that affected cash are summarized in the following
      table:

Cash paid for operating expenses	$210,000
Net increase in short-term debt	50,000

Cash paid for purchase of breeding stock	45,000
Cash paid for income taxes	25,000
Withdrawals	20,000
Repayment of term debt	22,000
Cash received from operations	220,000

a. Arrange the information in a statement of cash flows showing net cash provided by operating activities, investing activities, and financing activities as well as the net increase (decrease) in cash and the cash at the end of the year.

b. Putting yourself in the position of a lender providing operating credit to this cattle ranch for the upcoming year, discuss any concerns that you may have based on the information displayed in the statement of cash flows you have prepared.

■ Notes

1. Jack L. Smith, Robert M. Keith, and William L. Stephens, *Accounting Principles,* 3d ed. (McGraw-Hill, 1989), 1217–23.
2. The Farm Financial Standards Council, *Financial Guidelines for Agricultural Producers* (Farm Financial Standards Council, 1995), I-1.

6
Financial Statement Analysis

Liquidity Ratios
Solvency Ratios
Profitability Ratios
Repayment Capacity Ratios
Financial Efficiency Ratios
Loan-to-Value Ratio
Common-Size Statements and Horizontal Analysis
Questions and Problems

The information found on a set of financial statements can be made even more valuable using the analytical techniques discussed in this chapter. The primary technique that we will learn will be ratio analysis. A ratio, by definition, means that one number is divided by another. The answer, a quotient, can be expressed as a percentage (i.e., the numerator is a percentage of the denominator) or as so many dollars and cents of the numerator (x) for every one dollar of the denominator (i.e., x:1, read "x to one"). Ratios allow us to remove the size of businesses from influencing our analysis and theoretically compare the largest of agribusinesses to the smallest.

The majority of the ratios presented on the following pages are those recommended by the Farm Financial Standards Council.[1] In addition to the FFSC ratios, we will cover the loan-to-value ratio, which is commonly used by lenders. Realize that even more financial ratios than those found in this chapter exist and are used by investors, lenders, auditors, and owners. Whatever the ratio, one cannot even begin to understand its significance unless one knows the formula by which it is calculated.

The FFSC divides ratios into five categories depending on what is being measured: liquidity, solvency, profitability, repayment capacity, and financial efficiency. **Liquidity** is a short-term concept pertaining to the degree by which current assets exceed current liabilities. **Solvency** measures deal with the degree by which all assets exceed all liabilities. **Profitability** ratios put dollars of profits into relative terms by comparing profits to assets, equity, and gross revenue. **Repayment capacity** ratios are used in loan analysis in an attempt to predict whether or not a borrower will be able to pay principal and interest on loans as well as meet all other cash outflow obligations. Finally, **financial efficiency** ratios reflect the manager's ability to control costs and utilize assets efficiently.

■ Liquidity Ratios

The **current ratio** is a widely used measure of liquidity. Its calculation requires information from a single balance sheet—the current assets and current liabilities:

$$\text{Current ratio} = \frac{\text{Current assets}}{\text{Current liabilities}} \tag{6.1}$$

The answer gives an indication of the firm's bill-paying ability over the next twelve months. As of December 31, 1998, All American Farms had current assets of $228,000 and current liabilities of $76,000 (see table 5.3). Applying formula (6.1) yields a current ratio of 3:1. This means that All American Farms has $3.00 in current assets for every $1.00 in current liabilities. In other words, current assets are three times as large as current liabilities.

There is no universally established standard for the current ratio's minimum acceptable level. Certainly a ratio less than 1:1 would be alarming because it would mean that the firm's current assets are less than its current liabilities. In this case, the firm is illiquid and will probably not be able to meet its short-term obligations unless something is done. Even a ratio exactly equal to 1:1 would be of concern because there is no liquidity cushion. In a 1993 survey of California agricultural lenders, the modal minimally acceptable standard was that the current ratio should be greater than or equal to 1.25:1.[2]

The liquidity position of an agribusiness can depend on the time of the year and on the type of the business. Grain farms that are expending cash during the growing season will show a lower current ratio at that time than after harvest. Dairy farms are more likely to have a steady current ratio throughout the year since they continuously produce and sell milk and receive payments on a monthly or semimonthly basis. However, even dairy farms sometimes buy hay ahead in the summer months, which is likely to lower their current ratio at that time. Fruit and vegetable canneries experience large outflows of funds during the processing season, called the campaign, and then receive funds throughout the rest of the year. The financial analyst should take these industry and seasonal variations into account when assessing the liquidity of agricultural firms.

It is also prudent to look beyond the current ratio and consider the relative size of individual accounts within current assets and current liabilities. On the asset side, cash is completely liquid, but inventories and accounts receivable may require some time before they are converted into the cash needed to pay bills. Not all inventories are created equal; some are much more marketable than others. Accounts receivable can also include some potential bad debts, so it is best to conduct an aging of the accounts. Prepaid expenses and cash in growing crops, even though classified as current assets, are the least liquid. Prepaid expenses are never actually converted into cash; they only represent cash savings in that payments for these items will not have to be made within the next year. The conversion of the cash-in-growing-crops account into cash depends on how close the crops are to harvest and on what will be done with the crops at the time of harvest.

On the liability side, accounts payable require the use of cash in the near future while short-term notes payable and the current portion of term debt may not be paid for several months into the coming year. Similarly, accrued expenses, income-taxes payable, and the current portion of deferred taxes will vary in the timing of their payment due dates compared to the date of the balance sheet. In addition to an examination of the accounts affecting the current ratio, a more thorough analysis of the future liquidity of a firm would be to prepare a cash-flow budget as was presented in chapter 3.

Working capital is an alternative measure of liquidity to the current ratio. The calculation of working capital does not result in a ratio, but in an absolute amount of dollars:

$$\text{Working capital} = \text{current assets} - \text{current liabilities} \qquad (6.2)$$

As of December 31, 1998, the working capital of All American Farms was equal to $152,000, which, using equation (6.2), is $228,000 in current assets minus $76,000 in current liabilities. If All American Farms were to pay all its current liabilities with the cash generated by its current assets, it would have $152,000 left over as a cushion.

In general, the higher the working capital of a firm the better. However, too much working capital (and a current ratio that is very high) could mean that the firm is sitting on excess cash, uncollectable accounts receivable, or antiquated inventory. Excess cash could be put to work in buying productive assets that should generate a return higher than the interest the cash is earning in a bank. Accounts receivable and inventory should both be managed so that they are turning over at a rate that is reasonable for the type of industry in which the firm operates.

Working capital should not be compared between firms of different sizes. Two firms with the same current ratio could have vastly different amounts of working capital. For example, All American Farms has a current ratio of 3:1 and working capital of $152,000. A larger farm may have current assets of $3,000,000 and current liabilities of $1,000,000. The current ratio of this second farm would also be 3:1, but its working capital would be $2,000,000— much more than $152,000. Working capital is a very good indicator of liquidity, but only when looking at one firm at a time.

■ Solvency Ratios

There are three interrelated ratios that measure solvency—the debt-to-equity ratio, the debt-to-asset ratio, and the equity-to-asset ratio. Solvency is the ability to pay off all debts upon the liquidation of a firm. Liquidation is the selling of all the assets of the firm for cash. Unfortunately, the word "liquidation" sounds a lot like liquidity, and students often confuse the concepts of solvency and liquidity. They are not the same. While liquidity focuses on the short term, solvency is a long-term concern. Solvency looks at all assets and liabilities, not just current assets and liabilities. A firm is technically solvent if its total assets exceed its total liabilities and insolvent if its total assets are less than its total liabilities so that its owner's equity is a negative number.

All the information needed to calculate the three solvency ratios comes from the balance sheet. The names of the ratios describe their formulas:

$$\text{Debt-to-equity ratio} = \frac{\text{Total debt}}{\text{Total owner's equity}} \qquad (6.3)$$

$$\text{Debt-to-asset ratio} = \frac{\text{Total debt}}{\text{Total assets}} \qquad (6.4)$$

$$\text{Equity-to-asset ratio} = \frac{\text{Total owner's equity}}{\text{Total assets}} \qquad (6.5)$$

The balance sheet in table 5.3 shows that All American Farms had total assets of $865,000, total liabilities of $246,000, and total owner's equity of $619,000 at the end of 1998. Using formulas (6.3), (6.4), and (6.5), we calculate the solvency ratios to be:

$$\text{Debt-to-equity ratio} = \frac{\$246,000}{\$619,000} = 0.397:1$$

$$\text{Debt-to-asset ratio} = \frac{\$246,000}{\$865,000} = 0.284:1$$

$$\text{Equity-to-asset ratio} = \frac{\$619,000}{\$865,000} = 0.716:1$$

These three ratios can be interpreted as 39.7 cents of debt for every dollar of equity, 28.4 cents of debt for every dollar of assets, and 71.6 cents of equity for every dollar of assets.

Since the three solvency ratios are interrelated, we can use just one of the three to calculate the other two. Obviously, the debt-to-asset and equity-to-asset ratios will always be complements (i.e., their sum will be 1.0). But how does one find the debt-to-equity ratio from the debt-to-asset or equity-to-asset ratios? How does one go in the opposite direction to find the debt-to-equity ratio from either of the others? The answer is to build a very small balance sheet from the information that one ratio gives. As an example, say that we knew only that an agribusiness has a debt-to-equity ratio of 0.60:1. On the corresponding very small balance sheet, owner's equity is $1.00 and debt (total liabilities) is $0.60. Since total assets equal total liabilities plus owner's equity, the corresponding total assets must be $1.60. The debt-to-asset ratio is then $0.60 in debt divided by $1.60 in assets or 0.375:1. The equity-to-asset ratio is $1.00 in equity divided by $1.60 in assets or 0.625:1.

The debt-to-equity ratio is often called the leverage ratio. **Leverage** is the degree to which a business is financed by debt instead of equity. The higher the debt-to-equity ratio of an agribusiness, the higher its financial leverage, and the closer the business is to being insolvent. The equity-to-asset ratio is sometimes called the capitalization ratio, for it tells to what degree the firm is capitalized by equity.

Lenders are very concerned with the solvency ratios of their borrowers. In the California study previously cited, the modal response for the lenders interviewed was that the debt-to-equity ratio for a farm or ranch borrower should not exceed 1:1.[3] This would correspond to a debt-to-asset ratio not exceeding 0.5:1 and an equity-to-asset ratio not being less than 1:1. Yet, as with all the ratios presented in this chapter, please understand that there are no fixed standards or cutoff values by which a single ratio will be used to accept or deny a loan. All loans present an array of variables, which should be judged as a group.

A final point about solvency ratios is that the method used to value assets on the balance sheet can influence the results. Solvency ratios calculated on assets valued at appreciated market values will appear healthier than those calculated on assets valued at lower historical costs. The fact that solvency involves the ability of assets to exceed debts when *assets are liquidated* implies that market value, and not historical cost, should be used.

■ Profitability Ratios

Profits are reported in dollars on the income statement, and the maximization of profits is usually assumed by agricultural economists as the goal of the firm. Certainly, making a profit is a laudable achievement, but it is the size of that profit in relation to the assets owned and employed by a business that furnishes the best indication of management's achievement for the year. The ratio that the FFSC recommends to measure profitability relative to farm assets is the **rate of return on farm assets**, which is usually stated as an interest rate:

$$\text{Rate of return on farm assets} = \frac{\text{Net farm income from operations} + \text{Farm interest expense} - \text{Owner withdrawals for unpaid labor and management}}{\text{Average total farm assets}} \quad (6.6)$$

Notice that net farm income from operations (NFIFO) is used as the profit figure in formula (6.6). A review of the income statements in chapter 5 (tables 5.1 and 5.4) shows that NFIFO equals gross revenues minus operating expenses and interest expense. We do not use net farm income or net income as indicators of profits because they would include the somewhat arbitrary gain or loss on the sale of capital assets, miscellaneous revenue or expense, and income tax expense. These three items either are outside the normal, day-to-day running of the farm or are affected by variables external to the farm.

Farm-interest expense, which has been subtracted out in NFIFO, is then added back in as the second term in the numerator in formula (6.6). This is done because we want to measure the rate of return on farm assets *regardless of how those assets are financed.* Interest expense is added back in because it deals with how the assets were financed; it is a charge for debt financing. Suppose we are using the rate of return on farm assets to compare the profitability between two farm businesses. Both of the farms have the same gross revenues and the same operating expenses, but one of the farms is highly leveraged and incurs a large interest expense. The second farm has no debt, so it pays no interest. To put the assets of the two farms on an equal footing with regard to financing, we utilize the preinterest return. Therefore, we must add interest expense back into NFIFO for the first of the two farms.

The third term in the numerator of formula (6.6) shows a subtraction for owner withdrawals for unpaid labor and management. This is also done so the rate of return on farm assets can be compared fairly between two or more farms. The income statements for farms and ranches organized as sole proprietorships and partnerships include no deduction for the labor and management provided by the owner(s). Any such remuneration must come out of owner withdrawals, which are not expenses. Additionally, the owner's family members may provide labor that goes unpaid. Farms and ranches organized as corporations are allowed to deduct wages and salaries paid to owners if those owners are employed by the corporation. In the ratio, an estimated expense for unpaid management and labor must be deducted for sole proprietorships and partnerships to make their ratios comparable with those of corporations. But how does one estimate this charge? It is based on the concept of an opportunity cost that is defined as the income forgone from choosing one course of action over the next-best alternative. In this situation, the course of action chosen is to work on the farm. The next-best alternative would be doing work of similar difficulty and responsibility for some other business. We therefore could estimate what the owner(s) and their family members would make in hypothetical alternative positions of employment. Withdrawals removed from the business by the owner up to this estimate should then be subtracted out in the numerator.

The denominator in formula (6.6) also requires a bit of explanation. Average total assets are specified instead of beginning or ending assets. The returns in the numerator are realized over a year's time on assets that are changing. It would be misleading to attribute all of the returns to beginning or to ending assets. As a compromise, we use the average: calculated as the sum of beginning and ending total assets divided by two. The valuation of assets, whether at historical cost or market value, influences the resulting rate of return on farm assets, just as it also influenced the solvency ratios.

For All American Farms in 1998, table 5.4 shows a net farm income from operations of $94,000. The accrual-adjusted income is used because it better represents performance for the year. Interest expense was $18,000. Total owner withdrawals of $40,000 are found on the statement of owner's equity (table 5.2) and statement of cash flows (table 5.5). We will assume that this amount is a valid estimate for unpaid labor and management of the owners and their family members. Total assets were $792,000 at the beginning of the year, and they grew to $865,000 at the end of the year. Substituting these figures into formula (6.6) yields the following:

$$\text{Rate of return on farm assets} = \frac{\$94,000 + \$18,000 - \$40,000}{\dfrac{\$792,000 + \$865,000}{2}} = 8.69\%$$

A rate of return on farm assets of 8.69% is pretty good for production agriculture. Often, rates of return can be low for farms and ranches because they require large amounts of high-valued assets in the form of land and equipment to return what might be called a "living wage." However, it should be pointed out that capital gains due primarily to real property appreciation are not included in calculating the rate of return on farm assets. When capital gains are included, the average rate of return on assets for all U.S. farms and ranches during the period 1991 to 1996 is reported to be 5.12%.[4] In contrast, a southwest Minnesota study encompassing the years 1960 to 1988 shows returns to be greater than the national average for the farms surveyed. Average rates of return were 10.7% for typical farms when capital gains were included and were 8.2% when only ordinary income was measured.[5]

The second profitability ratio, the **rate of return on farm equity**, has the following formula:

$$\text{Rate of return on farm equity} = \frac{\begin{array}{c}\text{Net farm income} \\ \text{from operations}\end{array} - \begin{array}{c}\text{Owner withdrawals for} \\ \text{unpaid labor and management}\end{array}}{\text{Average total farm equity}} \qquad (6.7)$$

The rate of return on farm equity is also expressed as a percentage. The two terms in the numerator have been explained previously. We do not add interest back in as we did with the rate of return on farm assets because returns to equity are residual returns after the cost of debt, that is, interest, has been paid. The denominator again utilizes an average, in this case the sum of beginning and ending total farm equity divided by two.

For All American Farms in 1998, NFIFO was $94,000 and owner withdrawals for unpaid labor and management was $40,000. The numerator for formula (6.7) is therefore equal to $54,000. From the balance sheet (table 5.3), beginning and ending total farm equity amounts were $567,000 and $619,000, respectively, making average total farm equity equal to $593,000. The resulting rate of return on farm equity is:

$$\text{Rate of return on farm equity} = \frac{\$94,000 - \$40,000}{\$593,000} = 9.1\%$$

The rate of return on farm equity of 9.1% for All American Farms is higher than the rate of return on farm assets of 8.69%. This is as it should be if All American Farms is using leverage effectively. Theoretically, the firm would be able to achieve a rate of return on assets of 8.69% without borrowing (remember that this rate is measured without regard to how the assets are financed). With no debt, this 8.69% would also be the rate of return on equity since total assets would equal total equity. All American Farms should not borrow unless it can increase the rate of return on equity above the 8.69% figure. And that's just what it was able to do in 1998. If a firm takes on debt and realizes a lower rate of return on equity than on assets, then why borrow? Well, in reality there are other longer-term reasons for borrowing. These reasons include borrowing to acquire sufficient assets to benefit from economies of scale, to generate enough income to support a family, or to take advantage of potential long-term capital gains. Additionally, we should not just examine one year's returns to determine whether a firm is using leverage effectively. Since returns vary from year to year, it's the rate of return on equity being greater than the rate of return on assets over the long term that really matters.

The third and last profitability ratio is the **operating-profit-margin ratio**. It relates profits from operations to sales, and, like the other two profitability ratios, is stated as a percentage:

Operating profit margin ratio =

$$\frac{\begin{array}{c}\text{Net farm income} \\ \text{from operations}\end{array} + \begin{array}{c}\text{Farm interest} \\ \text{expense}\end{array} - \begin{array}{c}\text{Owner withdrawals for} \\ \text{unpaid labor and management}\end{array}}{\text{Gross revenues}} \qquad (6.8)$$

This is the first of our ratios where no data are needed from the balance sheet. Net farm income from operations, farm interest expense, and gross revenues all are found on the income statement—preferably an accrual or accrual-adjusted income statement. Owner withdrawals for unpaid labor and management, as in calculating the rate of return on farm assets, must be estimated using an opportunity cost. The operating-profit-margin ratio tells us how many cents out of every dollar of revenues are earned in the form of preinterest operating profit. Again, farm interest expense is added back in so the result is not influenced by the source of financing. Owner withdrawals for unpaid labor and management are subtracted out so farms and ranches with all types of legal organization can be compared.

For All American Farms in 1998, NFIFO was $94,000, interest expense was $18,000, unpaid labor and management were estimated to be $40,000, and accrual-adjusted gross revenues were $461,000. The calculation of the operating-profit-margin ratio is as follows:

$$\begin{array}{c}\text{Operating profit} \\ \text{margin ratio}\end{array} = \frac{\$94,000 + \$18,000 - \$40,000}{\$461,000} = 15.62\%$$

There is no common standard for this ratio, although most analysts would say that the higher it is the better. The magnitude of the operating profit margin reflects the firm's ability to generate revenues and control costs.

■ Repayment Capacity Ratios

Whether or not a borrower will have enough funds in the future to pay the principal and interest that are due on loans is of special interest to lenders. Repayment capacity ratios, which are also called coverage ratios, are designed to indicate the potential default risk that lenders are undertaking when they grant a loan. The FFSC recommends the use of the **term debt and capital lease coverage ratio** and the **capital replacement and term debt repayment margin**. The second of these is not a ratio, but is measured as an amount of dollars. After these two FFSC measures are discussed, we will then introduce an approach that is commonly used in lending for gauging the repayment capacity on short-term loans.

The numerator of the term debt and capital lease coverage ratio consists of eight added or subtracted terms. This appears to be confusing, but the idea is a simple one: we are trying to convert an accrual or accrual-adjusted net farm income from operations into an estimate of the net cash available to pay principal and interest on term debt and capital leases. The denominator is simply the annual payments that will be made on term debt and capital leases. Obviously, the numerator should be greater than the denominator if the term debt and capital lease payments are going to be made.

Term debt and capital lease coverage ratio =

$$\frac{\begin{array}{l}\text{Net farm income from operations} \pm \text{Miscellaneous revenue or expense} \\ + \text{Nonfarm income} + \text{Depreciation expense} + \text{Interest on term debt} \\ + \text{Interest on capital leases} - \text{Income tax expense} - \text{Owner withdrawals}\end{array}}{\begin{array}{l}\text{Annual scheduled principal and interest payments on term debt} \\ + \text{Annual scheduled principal and interest payments on capital leases}\end{array}} \quad (6.9)$$

In the numerator in formula (6.9), the starting point is net farm income from operations taken from the income statement. It is recommended that an accrual or accrual-adjusted income statement be used so that the NFIFO includes changes in current assets and current liabilities. Miscellaneous revenues are then added in and miscellaneous expenses subtracted out based on the assumption that these will be recurring items and will be received and/or paid in cash. Since nonfarm income can also be an important source of cash with which to pay loans, it is added in. Examples of nonfarm income are the wages and salaries received by a spouse or by the farmer for off-farm employment, the profits made on nonfarm investments, or repetitive cash gifts made by relatives of the farmer or the spouse. Depreciation expense had been taken out as an expense in the calculation of NFIFO. However, depreciation expense is added back in since it is a noncash expense and we are adjusting NFIFO to be more cashlike. Interest on term debt and capital leases, though also subtracted out in arriving at NFIFO, is a cash cost. This interest is added back in because the numerator of the term debt and capital lease ratio is supposed to indicate total funds available before principal and interest on term debt and capital leases has been paid. Notice the interest added back in is not equal to total interest expense; it would not include any interest on short-term liabilities. Income tax expense and owner withdrawals are subtracted out because they are uses of cash that are not accounted for in NFIFO.

The denominator of formula (6.9) requires less explanation. It consists of only the coming year's payments of principal and interest on term debt and capital leases. These are just one year's payments, not the total liability taken from the balance sheet.

Not all the data needed to calculate the term debt and capital lease coverage ratio for All American Farms can be found on the financial statements in chapter

5. Supplementary information must be provided: nonfarm income for 1998 was $20,000; interest on term debt was $15,000; the firm had no capital leases and thus no interest (or principal) to pay on capital leases; the scheduled principal and interest payments on term debt for 1999 will amount to $21,000. All other figures but one in the following ratio calculation came from the accrual-adjusted income statement (table 5.4). The figure for owner withdrawals is found in the statement of owner's equity (table 5.2) or the statement of cash flows (table 5.5).

$$\text{Term debt and capital lease coverage ratio} = \frac{\$94,000 + \$1,000 + \$20,000 + \$52,000 + \$15,000 + \$0 - \$13,000 - \$40,000}{\$21,000 + \$0} =$$

$$\frac{\$129,000}{\$21,000} = 6.14:1$$

A term debt and capital lease coverage ratio of 6.14:1 is very good. It says that All American Farms' projected net cash available is more than six times its scheduled payments on term debt and capital leases. In the California study previously cited, the modal standard offered by a sample of agricultural loan officers was that a coverage ratio should be greater than 1.25:1.[6]

The capital replacement and term debt repayment margin also gives the lender a forecast concerning the borrower's ability to repay. The **margin** is basically the money left over after payments have been made on term debt and capital leases. The phrase "capital replacement" in the title of this financial measure refers to the fact that a primary use of these leftover funds is to replace worn-out capital assets. The formula for the capital replacement and term debt repayment margin contains a lot of the same information as the formula for the term debt and capital lease coverage ratio:

+ Net farm income from operations
+ Miscellaneous revenue
− Miscellaneous expense
+ Nonfarm income
+ Depreciation expense
− Income tax expense
− Owner withdrawals
_____
= Capital replacement and term-debt repayment capacity

− Payment on unpaid operating debt from a prior period
− Principal payments on current portion of term debt
− Principal payments on current portion of capital leases
− Total annual payments on personal liabilities
                    (if not included in withdrawals)
_____
= Capital replacement and term-debt repayment margin (6.10)

Formula (6.10) is divided into two parts. The upper part calculates the capital-replacement and term debt repayment capacity. The only difference between this figure and the numerator of the term debt and capital lease coverage ratio (formula 6.9) is that interest expense on term debt and capital leases is not added back into NFIFO in formula (6.10). This is so because interest expense is also not included in the term debt and capital lease payments found in the lower part of formula (6.10). Only principal payments are subtracted out. We would obtain the same final answer if we added interest expense back into the top and then subtracted it out in the bottom section, but that would be a bit redundant.

Also in the lower part of formula (6.10), notice that two new items are subtracted. The first is the payment on unpaid operating debt from a prior period. The key to this item is that the "prior period" refers not to the year of the financial statements from which the data are drawn (like 1998 for the All American Farms example) but to years prior to that (like 1997 or 1996 for All American). A farm may have an outstanding operating loan balance as of the end of its accounting year. We assume that this loan will normally be repaid from funds flowing into the business. However, if any principal is still owed from previous years, then repayment of these unpaid carryovers will entail a special burden to the firm and should be accounted for. The other new item is the payment on personal liabilities. Often farm families have personal debts, such as home mortgages, that will require the use of funds. These should also be subtracted out if they are not already included in the owner withdrawals found in the upper part of formula (6.10).

For All American Farms, calculation of the capital replacement and term debt repayment margin requires additional data to that given above. At the end of 1998, All American had no unpaid operating loans from a prior period and all personal liabilities had been included in the $40,000 owner withdrawal. The principal portion of the term debt payment is $6,000 (i.e., $21,000 total payment minus $15,000 interest).

+ NFIFO	$94,000
+ Misc. revenue	+ 1,000
− Misc. expense	−0
+ Nonfarm income	+20,000
+ Depreciation expense	+52,000
− Income tax expense	−13,000
− Owner withdrawals	−40,000
= CR and TD repayment capacity	$114,000
− Prior period unpaid operating debt	$ 0
− Principal payments on term debt	−6,000
− Principal payments on capital leases	−0
− Payments on personal liabilities	−0
= CR and TD repayment margin	$108,000

Since the capital replacement and term debt repayment margin is a dollar amount, it is impossible to compare it for farms and ranches of different sizes. The lender certainly would require that it be positive, and the more positive the better. All American Farms' margin of $108,000 is very healthy, and it is in line with the term debt and capital lease coverage ratio of 6.14:1.

The term debt and capital lease coverage ratio and the capital replacement and term debt repayment margin are based on the assumption that history repeats itself. Many of the elements in both measures are taken from a single historical income statement, so we are using limited past information to predict repayment capacity in the future. If income is volatile, as it often can be in agriculture, then one year's performance may not be a valid predictor of the following year's. An alternative would be to base coverage and margin calculations on average data from the firm for the past several years in cases where that information is available.

Another device often used for checking on a firm's future ability to service debt is the cash-flow budget that was developed in chapter 3. The cash-flow budget is a projection so it avoids the problem of naively assuming that next year will be a carbon copy of last year. The annual cash-flow budget that appears in table 6.1 contains a summary column, which can be used for repayment capacity analysis.

The summary column of a cash-flow budget treats the entire year as a single period. For the cash receipts and payment entries, the summary column is the summation of the corresponding twelve monthly entries. All American Farms is projected to have cash receipts of $480,000 and to make cash payments of $461,000 in 1999. The net cash flow is projected to be $19,000 or $480,000 − $461,000. The beginning cash balance of $122,000 in the summary column is not a summation, for it makes no sense to sum stock items such as cash balances. We simply use the ending cash balance for 1998, which is taken from the 1998 balance sheet. Similarly, the ending cash balance of $134,400 is the ending balance from the December column, that is, the ending balance for the whole year.

From a repayment capacity point of view, the ending balance of $134,400 can also be thought of as the projected margin the borrower would have on the short-term line of credit. To calculate a coverage ratio, we know from the summary column that the borrower is projected to pay $165,400 in principal and $6,600 in interest on the line of credit. These combine to be the denominator in the coverage ratio. The numerator should reflect total cash available, and this is equal to funds that cover the principal and interest payments plus the excess cash represented by the ending balance.

$$\frac{\text{Line of credit}}{\text{coverage ratio}} = \frac{\$165,400 + \$6,600 + \$134,400}{\$165,400 + \$6,600} = \frac{\$306,400}{\$172,000} = 1.78:1$$

Table 6.1. All American Farms cash-flow budget for 1999

	January	February	March	April	May
Cash receipts					
Soybean sales		$30,000			
Corn sales	$ 30,000				
Calf sales					
Collection of receivables	17,000				
Other farm income					$ 3,000
Nonfarm income	1,600	1,600	$ 1,600	$ 1,600	1,600
Total cash receipts	48,600	31,600	1,600	1,600	4,600
Cash payments					
Rent					25,000
Labor and benefits	4,000	4,000	4,000	4,000	4,000
Fertilizer				36,000	
Seed					40,000
Chemicals					12,000
Repairs and maintenance	1,500	1,500	1,500	1,500	3,000
Fuel, oil, and grease	500	500	500	500	2,000
Utilities	500	500	500	500	500
Purchased feed		5,000		5,000	
Veterinary and medicine		1,000			1,000
Property taxes				3,000	
Office expenses	500	500	500	500	500
Accounts payable	35,000				
Short-term notes payable	26,000				
Accrued interest	3,000				
Payment of term debt	21,000				
Income taxes			5,000		
Owner withdrawals	3,500	3,500	3,500	3,500	3,500
Total cash payments	95,500	16,500	15,500	54,500	91,500
Net cash flow	(46,900)	15,100	(13,900)	(52,900)	(86,900)
Plus beginning cash	122,000	75,100	90,200	76,300	23,400
Cash without borrowing	75,100	90,200	76,300	23,400	(63,500)
Current borrowing					73,500
Principal repayment					
Cumulative borrowing					73,500
Interest payment					
Ending cash balance	75,100	90,200	76,300	23,400	10,000

June	July	August	September	October	November	December	Summary
				$140,000	$30,000		$200,000
				60,000		$ 75,000	165,000
			$ 75,000				75,000
							17,000
							3,000
$ 2,000	$ 2,000	$ 1,600	1,600	1,600	1,600	1,600	20,000
2,000	2,000	1,600	76,600	201,600	31,600	76,600	480,000
					25,000		50,000
4,000	4,000	4,000	4,000	4,000	4,000	4,000	48,000
14,000							50,000
							40,000
12,000	12,000						36,000
3,000	3,000	3,000	3,000	1,500	1,500	1,500	25,500
2,000	2,000	2,000	2,000	500	500	500	13,500
500	500	500	500	500	500	500	6,000
	10,000						20,000
		1,000			1,000		4,000
						3,000	6,000
500	500	500	500	500	500	500	6,000
							35,000
							26,000
							3,000
							21,000
8,000			8,000			8,000	29,000
3,500	3,500	3,500	3,500	3,500	3,500	3,500	42,000
47,500	35,500	14,500	21,500	10,500	36,500	21,500	461,000
(45,500)	(33,500)	(12,900)	55,100	191,100	(4,900)	55,100	19,000
10,000	10,000	10,000	10,000	10,000	84,200	79,300	122,000
(35,500)	(23,500)	(2,900)	65,100	201,100	79,300	134,400	
45,500	33,500	12,900					165,400
			49,500	115,900			165,400
119,000	152,500	165,400	115,900				
			5,600	1,000			6,600
10,000	10,000	10,000	10,000	84,200	79,300	134,400	134,400

A coverage ratio of 1.78:1 is above the common standard of 1.25:1, and the margin of $134,400 is reassuring. Table 6.1 does show that All American Farms is able to pay off its line of credit before the end of the year. However, a lender would notice that the projected net cash flow is only $19,000 for the year, and a lot of the repayment capacity comes from a strong beginning cash balance. Without the $122,000 in beginning cash, the margin and coverage ratio would not be nearly as healthy. It should be noted that these line of credit margin and coverage ratio techniques based on the cash-flow budget are not part of the recommendations of the Farm Financial Standards Council.

■ Financial Efficiency Ratios

The FFSC recommends five financial efficiency ratios. These ratios are of greater use to the manager of a business than they would be to lenders. The first is the asset-turnover ratio, which measures how many times a year assets regenerate their dollar value in the form of sales. As with other ratios that involve assets in their calculation, the valuation of those assets will affect the final outcome. The answer to the asset-turnover ratio is best labeled as "times per year."

$$\text{Asset-turnover ratio} = \frac{\text{Gross revenues}}{\text{Average total assets}} \qquad (6.11)$$

A firm likes to see a high asset turnover since that means its assets are working more efficiently. Unfortunately, enterprises in production agriculture tend to have very low asset-turnover ratios, usually less than one time per year. The primary reason is that farms and ranches require large investments in non-current assets such as land and equipment. A produce broker, in contrast, would generate high sales from a relatively small investment in assets and have a high asset turnover.

Using formula (6.11), All American Farms' asset turnover for 1998 would be its accrual-adjusted gross revenues of $461,000 divided by average total assets of $828,500. The result is that All American's assets are turned over 0.556 times per year. Even though this doesn't look very good, some farms and ranches have asset turnovers that are even lower.

The asset turnover is related to profitability. A firm puts its assets to work to generate sales, and, out of those sales, costs are subtracted to arrive at net farm income from operations. This process is illustrated by the fact that the rate of return on farm assets (a profitability ratio) is equal to the asset turnover times the operating profit margin ratio:

$$\frac{\text{Rate of return}}{\text{on farm assets}} = \left[\frac{\text{Gross revenues}}{\text{Average total assets}}\right] \times$$

$$\left[\frac{\text{NFIFO} + \dfrac{\text{Interest}}{\text{expense}} - \dfrac{\text{Owner withdrawals for}}{\text{unpaid labor and management}}}{\text{Gross revenues}}\right]$$

The gross revenues in the numerator of the first fraction cancels out with the gross revenue found in the denominator of the second fraction to produce formula (6.6):

$$\frac{\text{Rate of return}}{\text{on farm assets}} = \frac{\text{NFIFO} + \dfrac{\substack{\text{Farm} \\ \text{interest} \\ \text{expense}}}{} - \dfrac{\text{Owner withdrawals for}}{\text{unpaid labor and management}}}{\text{Average total assets}}$$

In 1998, All American Farms had an asset turnover of 0.556 times per year and an operating profit margin ratio of 15.62%. Multiplying these two together yields a rate of return on farm assets of 8.68%, which only differs from the 8.69% we calculated earlier due to rounding.

The other four financial efficiency ratios relate expense items and net farm income from operations to gross revenues. The results are usually expressed as percentages. These four ratios are (1) the operating-expense ratio, (2) the depreciation-expense ratio, (3) the interest-expense ratio, and (4) the net-farm-income-from-operations ratio. All the information in their formulas comes from the income statement:

$$\text{Operating-expense ratio} = \frac{\dfrac{\text{Total operating}}{\text{expenses}} - \dfrac{\text{Depreciation}}{\text{expense}}}{\text{Gross revenues}}$$

$$\text{Depreciation-expense ratio} = \frac{\text{Depreciation expense}}{\text{Gross revenues}}$$

$$\text{Interest-expense ratio} = \frac{\text{Total farm interest expense}}{\text{Gross revenues}}$$

$$\text{Net farm income from operations ratio} = \frac{\text{Net farm income from operations}}{\text{Gross revenues}}$$

Certainly, a manager wants to keep the operating-expense, depreciation-expense, and interest-expense ratios as low as possible. The net-farm-income-from-operations ratio should be as high as possible. It is similar to the operating-profit-margin ratio, and both can be referred to as "profit margins." The difference is that the operating-profit-margin ratio has interest expense added back in and owner withdrawals for unpaid labor and management sub-tracted out in the numerator.

The calculation of these four ratios for All American Farms for 1998 fol-lows. Notice that the sum of these four ratios is 100%, which points out that net farm income from operations plus interest expense plus depreciation expense plus all other operating expenses will be equal to gross revenue.

$$\text{Operating-expense ratio} = \frac{\$349,000 - \$52,000}{\$461,000} = 64.4\%$$

$$\text{Depreciation-expense ratio} = \frac{\$52,000}{\$461,000} = 11.3\%$$

$$\text{Interest-expense ratio} = \frac{\$18,000}{\$461,000} = 3.9\%$$

$$\text{Net-farm-income-from-operations ratio} = \frac{\$94,000}{\$461,000} = 20.4\%$$

■ The Loan-to-Value Ratio

Lenders use the loan-to-value ratio to gauge their risk exposure to decreases in the market value of collateral. A lender who is forced to foreclose and take possession of collateral wants to be able to sell that collateral for more than is owed against it. The lower the loan-to-value ratio, the greater the chances of the lender being protected. Stated conversely, a low loan-to-value ratio means the borrower has a large amount of his/her own money invested in the collateral. The formula for the loan-to-value ratio is:

$$\text{Loan-to-value ratio} = \frac{\text{Original loan principal}}{\text{Appraised value of collateral}} \qquad (6.12)$$

The data for formula (6.12) do not come directly from any of the financial statements discussed in chapter 5. The loan-to-value ratio is calculated for one loan at a time, not using the balance sheet's summarized information for all loans. Strictly speaking, this ratio should probably not be included in a chapter on financial statement analysis, but it is used widely in loan underwriting along with some of the other ratios we have covered in this chapter.

To illustrate the loan-to-value ratio, assume that a farmer wants to buy a neighboring farm that is currently on the market. The negotiated selling price is

$750,000. The purchaser has $250,000 for a down payment and hopes to borrow the remainder from his or her lender. The lender has had the farm appraised, and the appraisal has verified that the property is worth $750,000. The lender and farmer are considering a $500,000 loan with a term of twenty-five years and a fixed interest rate of 9%. The loan-to-value ratio would be:

$$\text{Loan-to-value ratio} = \frac{\$500,000}{\$750,000} = 66.67\%$$

Notice that this ratio is stated as a percentage. It is the complement of the down-payment percentage. In this example, the farmer is making a 33.33% down payment.

The survey of California lenders found that a majority of the respondents stated that the loan-to-value ratio should be less than 65%.[7] This appears to be quite conservative to the authors of this text. Even the above example with what seems like a relatively large down payment of $250,000 would miss qualifying by a little. Lenders may be willing to relax a low loan-to-value ratio standard in some cases depending on such factors as the borrower's history, the productivity of the collateral, the soundness of other financial ratios, and the prevailing economic outlook. Another technique that could be employed is to require secondary collateral, if available, which would strengthen the lender's position.

■ Common-Size Statements and Horizontal Analysis

On a common-size financial statement, the individual accounts are displayed as percentages. This exercise is most often applied to the income statement and the balance sheet. On the common-size income statement, revenue items, expenses, and measures of income are expressed as percentages of gross revenues. The common-size balance sheet will show asset, liability, and owner's equity accounts as percentages of total assets (which is the same as the percentage of total liabilities plus owner's equity). Such statements allow an analyst to compare companies of various sizes or to compare the statements of the same company over time. Tables 6.2 and 6.3 show All American Farms' accrual-adjusted income statement and balance sheet originally seen in chapter 5 in common-size format.

Another method used to analyze financial statements is horizontal analysis, also known as trend analysis. This refers to comparing the financial statements of a single company for more than one year. In this way, the changes from year to year can be identified, and those changes are often put in percentage form. Trends can be observed in specific financial variables such as sales, net income, total assets, and owner's equity. Table 6.4 shows an example of horizontal analysis as it applies to All American Farms for the last five years. Certainly more years and more financial variables can be included.

Table 6.2. All American Farms' common-size accrual-adjusted income statement for the year ended December 31, 1998

		(%)
Soybean sales	$196,000	42.5
Corn sales	183,000	39.7
+ Increase in crop inventories	10,000	2.2
Calf sales	72,000	15.6
− Decrease in accounts receivable	(5,000)	(1.1)
Other farm income	5,000	1.1
Gross revenues	$461,000	100.0
Rent	$50,000	10.8
Fertilizer	48,000	10.4
Labor and benefits	45,000	9.8
Seed	37,000	8.0
Chemicals	33,000	7.2
Repairs and maintenance	26,000	5.6
Fuel, oil, and grease	14,000	3.0
Purchased feed	22,000	4.8
Veterinary and medicine	3,000	0.1
Utilities	6,000	1.3
Depreciation expense	52,000	11.3
Property taxes	5,000	1.1
Office expense	3,000	0.7
+ Increase in accounts payable	3,000	0.7
+ Decrease in prepaid expenses	2,000	0.4
Operating expenses	349,000	75.7
Interest paid	17,000	3.7
+ Increase in accrued interest	1,000	0.2
Interest expense	18,000	3.9
Total expenses	367,000	79.6
Net farm income from operations	94,000	20.4
Gain (loss) on the sale of capital assets	10,000	2.2
Net farm income	104,000	22.6
Miscellaneous revenue (expense)	1,000	0.2
Income before income taxes	105,000	22.8
Income taxes paid	14,000	3.0
Decrease in current portion of deferred taxes	(1,000)	(0.2)
Income tax expense	13,000	2.8
Net income	$ 92,000	20.0

Table 6.4 shows that All American Farms has prospered over the last five years, showing positive net farm income from operations and increases in owner's equity in each year. The percentage changes in table 6.4 use the beginning year as the base on which the percentage change is calculated. For example, between 1994 and 1995, sales (gross revenues) increased from $322,000 to $345,000. This is an absolute change of $23,000, and $23,000 divided by 1994 sales of $322,000 equals the positive 7% change that appears in the table. Of course, when the base on which the percentage change is calculated is relatively small as it is with NFIFO, the percentage changes will be larger. In fact, the percentage changes in NFIFO in table 6.4 point out that the NFIFO, although positive during all five years, is quite volatile.

Table 6.3. All American Farms' common-size balance sheet for the year ending
December 31, 1998

		(%)
Assets		
Cash	$122,000	14.10
Inventory	60,000	6.94
Accounts receivable	17,000	1.97
Prepaid expenses	4,000	0.46
Cash in growing crops	25,000	2.89
Total Current Assets	228,000	26.36
Breeding livestock	150,000	17.34
Less: accumulated depreciation	(42,000)	(4.86)
Machinery and equipment	360,000	41.62
Less: accumulated depreciation	(159,000)	(18.38)
Buildings and improvements	160,000	18.50
Less: accumulated depreciation	(72,000)	(8.32)
Land	240,000	27.75
Total Noncurrent Assets	637,000	73.64
Total Farm Business Assets	$865,000	100.00
Liabilities		
Accounts payable	$35,000	4.05
Short-term notes payable	26,000	3.01
Current portion of term debt	6,000	0.69
Accrued interest	3,000	0.35
Income taxes payable	5,000	0.58
Current portion of deferred taxes	1,000	0.12
Total Current Liabilities	76,000	8.79
Noncurrent portion of notes payable	32,000	3.70
Noncurrent portion of real estate debt	111,000	12.83
Noncurrent portion of deferred taxes	27,000	3.12
Total Noncurrent Liabilities	170,000	19.65
Total Liabilities	246,000	28.44
Retained capital	556,000	64.28
Valuation equity	63,000	7.28
Total Owner's Equity	619,000	71.56
Total Liabilities and Owner's Equity	$865,000	100.00

Table 6.4. Horizontal analysis of All American Farms

	1994	1995	1996	1997	1998
Sales	$322,000	$345,000	$438,000	$402,000	$461,000
Percentage change	+7%	+27%	−8%	+15%	
Operating expenses	$259,000	$264,000	$337,000	$321,000	$349,000
Percentage change	+2%	+28%	−5%	+9%	
NFIFO	$48,000	$67,000	$86,000	$64,000	$94,000
Percentage change	+40%	+28%	−26%	+47%	
Total assets	$703,000	$723,000	$766,000	$792,000	$865,000
Percentage change	+3%	+6%	+4%	+9%	
Total liabilities	$258,000	$244,000	$236,000	$225,000	$246,000
Percentage change	−5%	−3%	−5%	+9%	
Owner's equity	$445,000	$479,000	$530,000	$567,000	$619,000
Percentage change	+8%	+11%	+7%	+9%	

■ Questions and Problems

6-1. Answer the following questions about three agribusinesses.

   a. An agribusiness has a debt-to-equity ratio of 0.44:1. What is its corresponding debt-to-asset ratio?
   b. Another agribusiness has a debt-to-asset ratio of 0.2:1. What is its corresponding equity-to-asset ratio?
   c. A third agribusiness has a debt-to-asset ratio of 0.85:1. What is its corresponding debt-to-equity ratio?
   d. Which of these three firms (a, b, or c) is most leveraged? Explain how you know.

6-2. Answer these asset-turnover questions about three firms.

   a. If a firm has an asset turnover of 0.6 times per year and an operating profit margin of 22%, what is its rate of return on farm assets?
   b. If a firm has a rate of return on farm assets of 4.2% and an operating profit margin of 17.5%, what is its asset turnover?
   c. If a firm has an asset turnover of two times per year and a rate of return on farm assets of 7.2%, what is its operating profit margin?

6-3. A greenhouse has an operating-expense ratio of 72%, a depreciation-expense ratio of 10%, and a net-farm-income-from-operations ratio of 13%. What is the greenhouse's interest-expense ratio?

6-4. A cash-flow budget of a winery projects that the ending cash balance for the upcoming year will be $114,000. Projected line of credit principal and interest payments are $608,000 and $36,000, respectively. Calculate the line of credit coverage ratio.

6-5. The accountant for Johnson Farms has prepared the following accrual-adjusted income statement (table 6.5) and balance sheets (table 6.6). See pages 108 and 109 for tables 6.5 and 6.6.

Supplementary information: (1) Owner withdrawals for unpaid labor and management were equal to $19,000 during 1999. (2) Annual scheduled principal and interest payments on term debt were $102,000 for principal and $72,000 for interest during 1999. (3) Johnson Farms had no nonfarm income, no capital leases, no payments on unpaid operating debt from a prior period, and no payments on personal liabilities during 1999.

   a. Using the income statement and balance sheets for Johnson Farms, calculate the ratios and measures listed below for 1999. Set up the formulas and show all the numbers you use in your calculations: current ratio, working capital, debt-to-equity ratio, debt-to-asset ratio, equity-to-asset ratio, rate of return on farm assets, rate of return on farm equity, operating-profit-margin-ratio, term debt and capital lease coverage ratio, capital replacement and term debt repayment margin, asset turnover, operating-expense ratio, depreciation-expense ratio, interest-expense ratio, and the net-farm-income-from-operations ratio.

Table 6.5. Johnson Farms' accrual-adjusted income statement for the year ended
December 31, 1999

Cattle sales	$566,000	
Hog sales	217,000	
Corn sales	185,000	
Government program payments	27,000	
Custom work	33,000	
+ Increase in accounts receivable	45,000	
− Decrease in inventory	(31,000)	
Gross Revenues		$1,042,000
Feed expense	$187,000	
Purchased market livestock	352,000	
Labor expense	115,000	
Seed	44,000	
Chemicals	26,000	
Fuel	19,000	
Utilities	16,000	
Repairs	5,000	
Veterinary and medicine	11,000	
Rent expense	42,000	
Property taxes	8,000	
Depreciation expense	56,000	
+ Increase in accounts payable	4,000	
Operating Expenses		885,000
Interest expense	140,000	
Total Expenses		1,025,000
Net Farm Income from Operations		17,000
Gain (loss) on the sale of capital assets		2,000
Net Farm Income		19,000
Miscellaneous revenue (expense)		(3,000)
Income Before Income Taxes		16,000
Income taxes paid	6,000	
+ Increase in current portion of deferred taxes	1,000	
Income tax expense		7,000
Net Income		$ 9,000

b. Write a two-page (or longer) essay summarizing and analyzing the financial ratios of Johnson Farms from an ag lender's perspective. What are the farm's good points? What are some problem areas? Be sure to address the issues of liquidity, solvency, profitability, repayment capacity, and financial efficiency in your essay.

6-6. Construct a common-size income statement and a common-size balance sheet for 1999 from tables 6.5 and 6.6.

Table 6.6. Johnson Farms' balance sheet as of December 31, 1998, and December 31, 1999

	1998	1999
Assets		
Cash	$ 5,000	$ 2,000
Accounts receivable	148,000	193,000
Inventory	290,000	259,000
Total current assets	443,000	454,000
Breeding livestock	95,000	97,000
Less: accumulated depreciation	(14,000)	(18,000)
Equipment	620,000	610,000
Less: accumulated depreciation	(124,000)	(167,000)
Buildings	384,000	384,000
Less: accumulated depreciation	(33,000)	(42,000)
Land	575,000	575,000
Total noncurrent assets	1,503,000	1,439,000
Total farm assets	$1,946,000	$1,893,000
Liabilities		
Accounts payable	$58,000	$62,000
Short-term notes payable	423,000	477,000
Current portion of term debt	102,000	102,000
Current portion of deferred taxes	25,000	26,000
Total current liabilities	608,000	667,000
Noncurrent portion of term debt	967,000	865,000
Deferred taxes—noncurrent	16,000	16,000
Total noncurrent liabilities	983,000	881,000
Total liabilities	1,591,000	1,548,000
Retained capital	335,000	325,000
Valuation equity	20,000	20,000
Total owner's equity	355,000	345,000
Total liability and owner's equity	$1,946,000	$1,893,000

■ Notes

1. Farm Financial Standards Council, *Financial Guidelines for Agricultural Producers* (The Farm Financial Standards Council, 1995), III-1–III-21.
2. Robert Thompson and Steven Blank, "Ag Loan Analysis: Criteria Used by Loan Officers in California," *Journal of Agricultural Lending,* winter 1994, 14.
3. Thompson and Blank, "Ag Loan Analysis."
4. U.S. Department of Agriculture, Economic Research Service, *Agricultural Income and Financial Situation and Outlook Report,* September 1996.
5. James Monke, Michael Boehlje, and Glenn Pederson, "Farm Returns: They Measure Up to Returns to Other Investments," *Choices,* first quarter 1992.
6. Thompson and Blank, "Ag Loan Analysis."
7. Thompson and Blank, "Ag Loan Analysis."

7
Risk in Agribusiness

Life is fraught with risk. We all face many risks in our personal and professional lives—catching a flu bug that is going around, being involved in a traffic accident, going through a divorce, losing a job, etc. Entrepreneurs take on even greater levels of risk. If the business goes under, the entrepreneur may lose a job along with losing the investment.

■ Financial and Business Risk

Businesses face two broad categories of risk—financial risk and business risk. **Financial risk** deals with the debt and equity structure of the business. The greater the leverage and the greater the debt-to-equity ratio for a firm, the higher its financial risk. A high level of debt requires that a firm must make high principal and interest payments. Therefore, the firm must generate annual cash flows that are sufficient to cover these principal and interest payments as well as all other cash obligations. It takes a lot of luck to carry out this high-wire act year after year. When a bad year hits, the firm is unable to service its debt, and creditors must be willing to postpone payment until the next year if the firm is going to survive. When bad years come in succession, the firm with a high debt load is unable to borrow any more funds, that is, it has no credit reserves. Creditors have little choice but to repossess and foreclose. Often under these circumstances, the firm is forced into bankruptcy.

Business risk is defined simply as the variation in returns (or net income) over time. In agriculture, it is often pointed out that returns can be volatile—high one year and low the next. Stated alternatively, agriculture shows a relatively high degree of business risk. The degree of business risk inherent in an agribusiness enterprise is often assumed away for budgeting purposes. The budgeter usually makes estimates for a "typical year," whatever that is. It is far more realistic to also consider the variation in returns from an enterprise, and this "what if" analysis has been made much easier with the advent of computer spreadsheets. Changes in the projected level of production, level of input use, and input and output prices can now be readily considered once the spreadsheet template has been constructed.

In discussing risk, we should not just dwell on the negative, meaning we should not just consider downside risk. Accepting a certain level of both financial and business risk can have its benefits. In chapter 6, we discussed the fact that leverage (i.e., taking on financial risk) is advantageous if the long-term rate of return on equity is greater than the long-term rate of return on assets. Business risk also has an upside. The fact that returns vary from year to year should result in returns that are higher than the average in some of the years. If we assume a normal distribution, then returns should be above the average in half of the years.

A **risk-free investment** is one that displays no business risk. A U.S. government bond is often given as an example of a risk-free investment. If an investor buys a one-year, one-thousand-dollar U.S. bond paying 5% interest, then the investor will receive the one thousand dollars in principal plus fifty dollars in interest at the end of a year without variation. However, most investments show some degree of business risk. In general, the astute investor should realize that for investments worthy of consideration, the higher the level of expected return, the higher the level of its business risk.

This direct relationship between risk and return is shown in figure 7.1. The variable on the horizontal axis is a measurement of the business risk of many possible investments or groups of investments called **portfolios**. Risk can be measured using the statistical concepts of variance, standard deviation, or coefficient of variation. All of these are beyond the scope of this book but are mentioned simply to emphasize that risk can be quantified. The variable on the vertical axis represents the returns from many possible investments. The return from an investment is assumed to be its average return, which is also called the expected return. Returns can be measured in dollars or in percentages, such as a rate of return on investment.

All points in figure 7.1 are the combinations of the risk and return from investments or investment portfolios. However, only some of the risk-return

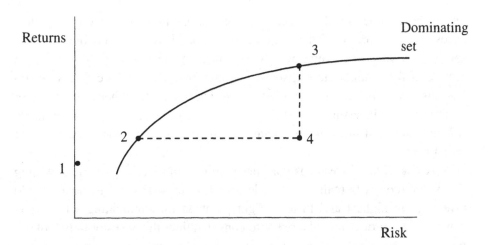

Figure 7.1. A dominating set.

combinations are available to the investor at a point in time. The risk-free investment is labeled as "1." Notice that it has a low return and does not travel out on the risk axis at all since it has zero risk. Investments 2 and 3 are on the **dominating (or efficient) set**. These are the investments that have minimum risk for a given level of return or maximum return for a given level of risk. To illustrate this concept, consider investment 4. Investment 4 is dominated by investment 2 because investment 2 has lower risk for the same return. Investment 4 is also dominated by investment 3 because investment 3 has higher return for the same level of risk. Investments 2 and 3 are two of the infinite number of possibilities on the dominating set, and there are no investments or portfolios currently available above the dominating set. Notice that the dominating set is upward sloping showing that, as has been pointed out before, increased returns can only be achieved with increased risk. Moreover, the dominating set is increasing at a decreasing rate because returns can only be increased by taking on greater and greater amounts of additional risk.

■ Attitudes toward Risk

The selection of investments by individuals will depend on what investments are available to them and the risk-return characteristics of those investments as shown in figure 7.1. A rational investor chooses either the risk-free investment or an investment along the dominating set. In the real world, the problem is identifying which investments are on the dominating set *before* the investment is undertaken.

Equally as important to a person's investment choice is the person's attitude toward risk. Rational investors are assumed to be **risk averse**, which means that they are only willing to take on more risk if they are rewarded with sufficiently higher returns. Each investor must decide just how much higher the returns have to be in order to accept more risk. Consequently, there are differences in the degrees of risk aversion between one investor and another. A person who is very risk averse will be satisfied on the lower risk–lower return section of the dominating set, such as at investment 2 in figure 7.1. A person who displays little risk aversion will be willing to accept higher risk for higher returns, as at investment 3 in figure 7.1. Risk aversion is not complete risk avoidance; it does not translate into all investors choosing the risk-free investment.

A risk-loving investor is considered to be irrational. He or she is willing to give up returns to realize greater levels of business risk. A risk lover would invest at the highest and furthest right point on the dominating set. A risk-neutral investor does not take risk into consideration; this investor only looks at returns. Risk-neutral investors try to maximize returns regardless of risk and would also invest at the highest point on the dominating set.

Our discussion has focused on someone called an investor. It should be pointed out that the ideas of the dominating set and risk aversion apply to agribusiness managers as well as to individuals making personal investments. The agribusiness manager must decide on what enterprises to undertake as well as what capital assets to purchase. These decisions involve a commitment of funds, the expectation of returns, and the acceptance of business risk just as does any investment.

■ Sources of Business Risk in Agriculture

There are a variety of sources from which business risk can arise in agriculture. Eight categories of risk will be covered in this section. These eight sources of risk are not mutually exclusive, meaning they can overlap and can be related one to another. By far, the two most prevalent are production risk and price risk.

Production risk refers to variations in a firm's dollar returns due to variations in the quantities of output it produces. In farming and ranching, many factors can contribute to production risk. Severe weather is the example that first comes to mind. Rain (or the lack thereof) determines the production for a dryland farmer or for a rancher running cattle on the open range. Excessive precipitation that leads to flooding has been known to wipe out all sorts of crops and livestock. Frost certainly can decimate a citrus or winter vegetable crop. Intense heat affects the quantity of milk produced by dairy cows. High winds can lodge a grain crop and reduce the amount of grain that is ultimately harvested.

Pests and diseases are also sources of production risk. Thousands of species of insects cause harm to both plants and animals. Vertebrate pests, such as gophers, birds, and deer, can damage crops, and predators can attack and kill livestock. Weeds are one of the costliest pests that farmers have to battle. Diseases that damage agricultural production run the gamut from mastitis in dairy cows to scleratina in iceberg lettuce.

Production risk can affect agribusinesses in the input supply and output marketing sectors as well as in the farm and ranch sector. A labor contractor would have little business if a frost has made crops unharvestable, or consider a winery that must cut production when the soil-borne disease phylloxera reduces the grape harvest.

Price risk, which is also known as market risk, is defined as the variation in returns due to changes in the prices of outputs and inputs. Cash markets for agricultural commodities tend to be quite volatile because of such factors as the structure of the industry, the production risk inherent in agriculture, domestic and international political developments, and the inelasticity of demand and supply. To the extent that production agriculture is purely competitive with many producers selling homogeneous products, farmers and ranchers are price takers. They operate at the mercy of seasonal patterns, cycles, trends, and externality induced episodic price movements. Changes in exchange rates further aggravate the situation making prices either higher or lower for our trading partners. Food processors also are exposed to price risks—when farmers are benefiting from high prices, food processors are suffering with higher costs, and vice versa.

The prices of inputs can vary too. During the energy crises of the 1970s, high prices for oil affected not only the cost of fuel to agribusinesses but also the costs of inorganic fertilizers, pesticides, and herbicides. Shifts in the supply of farm labor inversely affect wage rates. Changes in feed prices impact the livestock feeder and the farmer who raises the feed in opposite directions. The purchase prices of capital assets are also an important determinant of the overall profitability of the agribusiness. Finally, changes in interest rates are a form of price risk, since the interest rate is the price the firm pays for borrowed funds.

Production and price risk seem especially significant in agriculture. But the six other sources of business risk can be equally as beneficial or detrimental to an agribusiness firm at times. The risk of an accident, an injury, a fire, or an earthquake is known as **casualty risk**. Theft, the dishonesty of business associates, the resignation of employees, and the death of the owner are examples of **human risk**. **Legal risk** refers to the risk of being sued. **Political risk** includes any actions by domestic or foreign governments that impact one's business. Examples of political risk include changes in tax laws, trade laws, or regulations as well as declarations of trade embargoes, asset confiscation, and even war. **Economic risk**, the risk of inflation, recession, and depression, is closely associated with political risk to the extent that macroeconomic variables are a function of political decisions. Last, **obsolescence risk** involves the business manager adopting technologies that become outdated before they have paid for themselves. Certainly, there has been a lot of obsolescence risk in the purchase of personal computers and computer software over the last couple of decades.

■ Risk Management in Agriculture

From our discussion so far, we can see that financial risk and business risk are widespread in agribusiness. Is there anything a manager can do? Are there any ways that risks can be lowered?

Risk-management techniques will not eliminate risk, but they will lower the risk profile of the firm over the long run. Being a risk manager is simply part of being a good manager. A good manager is an astute planner and knows how to use records and financial statements. He or she can lead people to achieve a common goal, is abreast of the technologies that pertain to the company, is aware of marketing alternatives, has knowledge of the price movements for products, and pays attention to detail. All of these attributes of good management have tie-ins with risk management. However, in addition to these, there are techniques that a manager can use that specifically target financial and business risk.

Financial risk can be mitigated by maintaining **credit reserves**. Remember that credit reserves are defined as the ability to borrow more. Therefore, a firm with credit reserves is not too highly leveraged—possibly with a debt-to-equity ratio of less than 1:1. When bad years occur, a firm with credit reserves may be able to borrow enough to see it through until the economic situation improves. A related technique would be to **maintain savings**. With savings, the firm may not even have to increase debt during bad times, for it may be able to cover any negative cash flows by using its own funds.

Diversification is a technique that is often used by agribusinesses to reduce business risk. A farmer or rancher may combine two or more enterprises, or an agribusiness in the input supply or output marketing sectors may switch from being a single-product to a multiproduct firm. The idea is that in a year when one enterprise is doing badly, another enterprise may be doing well, and the net income for the total business will even out at an acceptable level. In a 1993 survey of California farmers and ranchers, diversification was reported to be the most widely used of eight possible risk management tools. Forty-eight percent of the respondents stated that they had used diversification to manage risk.[1]

For diversification to be most effective, all investments (enterprises) should be on the dominating set, and the returns from the investments (enterprises) should be negatively correlated. Yet, overall portfolio risk will be reduced even when the investments have positive correlation, as long as they are not perfectly positively correlated. The manager should realize that if diversification successfully reduces risk along the dominating set, then returns are also reduced. The agribusiness manager should also consider diversifying in enterprises that are nonagricultural. Moreover, diversification in agriculture is only applicable to the extent that a firm's resources can be adapted to several uses. A dairy barn has a rather specific use, as does a meat-packing plant, a vegetable cooler, or a hog-feeding facility. Investment in these types of large-ticket capital assets can dictate specialization and severely reduce the possibility of diversification by a business.

Insurance is an obvious method of compensating for losses arising from specific sources of business risk. Types of casualty and legal risk are often insured against. Fire insurance and liability insurance on real and personal property are common examples. Agribusinesses also maintain collision and liability insurance on cars and trucks used on public highways. Even life insurance is designed to mitigate the financial blow from the death of a family breadwinner.

Farmers can insure against specific sources of production risk with the use of crop insurance. There are two programs currently available. The one that covers against catastrophic loss is funded by the federal government and by payments from farmers. To take part in this program, a farmer must pay fifty dollars per crop per county plus a charge of 10% of the actuarial cost of the insurance. Claims are paid on losses greater than 50% of the producer's average yield using a price that is 55% of the expected market price. Any crop is insurable under this program, and farmers must participate in order to take part in other federal government programs, including disaster relief.

Under the second program, the farmer can increase coverage above the catastrophic level. This federally subsidized "buy-up" is available through private insurers and applies to about fifty different crops. The farmer is able to insure for a 25 to 50% loss in production. The smaller the percentage loss that is insured against, the higher the premium and the lower the government subsidy. The farmer is also able to set the price at which claims will be paid at between 60 and 100% of the price annually established by the Federal Crop Insurance Corporation.[2]

With **forward contracting**, an agricultural producer signs a contract with the buyer of the crop or livestock product, usually before production commences. The buyer agrees to purchase the product, and the producer agrees to sell to only that buyer. Many factors can be stipulated in these contracts, such as quantity, quality, production methods, postharvest handling techniques, timing, and method of payment. A very important factor that is often, though not always, stipulated in the contract is the price that the buyer will pay. With the establishment of a price in advance, the producer is able to reduce price risk. A related advantage of forward contracting to the producer is that the producer's product is guaranteed a home. The forward contract can reduce the production risk and price risk of the buyer; the buyer, which is often a food processor, has locked in the source of a basic raw ingredient and possibly the price that the buyer will pay for that ingredient.

Forward contracting is becoming more and more prevalent in agriculture with each passing year. Commodities for which forward contracts are commonly utilized include canning tomatoes, hogs, grains, wine grapes, processing potatoes, and crops grown for seed.

Another marketing technique used by farmers is **spreading sales** throughout the season in lieu of selling all of one's output at one time. Many agricultural product markets show seasonal price patterns with high prices at the beginning of the season, lower prices during the middle, and a possible upturn in prices as the season ends. By spreading sales, the farmer can average out the highs and lows in the seasonal pattern. If the product is storable, as with grains and dry beans, the farmer can market the product after everyone else's crop has been harvested. For perishable products, the farmer or rancher adjusts the planting schedule or other production practices to provide for several harvests and product sales throughout the season.

The final means of dealing with business risk that will be discussed is **hedging** techniques. A hedge is the process of taking a position in a security or asset equal to but opposite to one's existing position in another asset. For the commodity producer, say the corn grower, price risk is inherent during the growing season due to the farmer being a price taker as noted above. When the farmer begins planting, corn may be selling for, say, $2.00 a bushel. At that price, the grower can make a reasonable profit. Of course, the grower could forward contract (as noted above) with a food manufacturer and attempt to lock in this price. But additional flexibility can be gained by hedging using commodities futures contracts.

Since the mid-1800s, the Chicago Board of Trade (CBOT) has provided futures contracts to market participants as a means of reducing price risk. A **futures contract** is an agreement with the price set in advance to make delivery or take delivery of a quantity of a commodity at or before the contract expiration date. Because there is a delivery mechanism in place for trading at the Chicago Board of Trade, the futures contract has market value based on what price the real commodity is trading at on a given day. Although the contract price of corn futures may not exactly match the real or "cash market" price of corn, generally the two prices are very close.

The current pricing of corn futures at the CBOT is in cents per bushel, with prices quoted in whole numbers and fractions. For example, in July 1999, the December maturity corn futures contract is trading in the $196\frac{1}{2}$ range.[3] This would be read as $1.965 per bushel. Suppose our grower expected to harvest the crop in late October and deliver it by December to the buyer, a major food processor. If corn is running near $2.00 a bushel in April, and this is deemed a "good" price by the grower, then the grower can hedge by taking a position in the December corn futures contract by **selling short**. Shorting a futures contract is actually selling the obligation to deliver the commodity to the "long" position by the expiration date, but the price is set at the time of the short sale. So suppose the grower sells short one million bushels of corn April 1 at a price of 200 ($2.00 per bushel). Each futures contract requires five thousand bushels, so the grower needs to trade 200 contracts (200 = 1,000,000 ÷ 5,000). At a price of 200, this would mean each contract would be worth five thousand bushels times $2.00 or $10,000. If the grower sells short 200 contracts, the grower's account would be credited for 200 contracts times $10,000 or $2,000,000.

If, following the short sale on April 1, the value of corn declined due to the threat of a glut on the market, then the "cash" market price for corn would drop. Say the cash price dropped to $1.50 per bushel by December, when the grower delivers the crop. The grower no longer receives the profitable price of $2.00 per bushel, receiving $.50 less. The lost revenue for the total crop of one million bushels would be $500,000. But simultaneously, the grower's position in the futures contracts has *increased* by the same $500,000. In effect, the grower can buy out of the "short" position by paying the new value of the contract in December. Since the futures contract follows the cash market price for bushels of corn, it should drop to a quoted price of 150, which is $1.50 a bushel, or $7,500 per contract. Our grower initially sold short 200 contracts and received $2,000,000 credit to his or her account at the CBOT. Now the grower liquidates the trade by paying 200 contracts times $7,500 or $1,500,000, pocketing a $500,000 profit in the process. It is this profit on futures trading that exactly offsets the decline in revenue in the cash market, so the grower is back to where he or she started in April, in effect selling the corn crop for $2.00 a bushel.

What if instead of a bumper crop of corn, the market experienced drought, but our farmer was able to produce the one million bushels as he or she planned? Then the price per bushel would undoubtedly rise to, say, $3.00. Now the grower's cash crop could be sold in December for $3,000,000. But the grower didn't know what the season would hold for corn growers so the grower hedged as in the above case. This means the grower received the $2,000,000 credit to his or her account by selling short back in April, then is forced to liquidate the position in December. But like the cash price, the futures contract price should now be 300, which is $3.00 per bushel or $15,000 per contract. The grower has to buy out of the same 200 contracts so this costs $3,000,000 in December. Since the grower only received credit for $2,000,000 back in April, this is a $1,000,000 loss! So why hedge? Because the grower made an extra $1,000,000 from the sale of the corn in the cash market. So the two figures offset each other, and the grower still receives the original value of $2.00 per bushel that was an acceptable sales price to undertake raising the crop in the first place.

It is that aspect of hedging that troubles the new practitioner or student; that you can't keep the gains in either the cash market or in the futures contract trading, because these gains are always offset by losses in the other market. But that is okay. The hedge is designed to eliminate price risk, and by bringing the grower back to the originally acceptable price of $2.00 per bushel, it is effective.

Food processors also utilize futures contracts to hedge price risk. For example, suppose SunBlest Orange Juice is in continuous production and believes that a price of $.75 per pound will provide it with a reasonable profit for orange juice to be processed in May. The growing season begins in January and at that time the May futures contract for orange juice is trading at 75. Neither growers nor producers know how the orange crop will turn out— bumper or bust. So SunBlest goes long on orange juice futures contracts in sufficient quantity to hedge its price risk for its May processing. By going long, it has the right to the delivery of the contract's orange juice, so the "short" trader must liquidate his or her position at maturity or deliver. Suppose it is a high-frost year and a poor year for oranges. The cash market price for orange juice will rise. SunBlest will have to pay more for its purchase of oranges from growers, but will offset this additional cost by making a profit on its orange juice futures contract trading.

Here are some final comments about hedging using commodities futures contracts. First, it does not take a speculator to make the system work, only a party with the opposite interest. Initial trades require that the party (grower or processor) put up "margin money," or earnest money. The accounts are adjusted daily by the CBOT, and if the party has made a profit, it is credited to the party's account; for a loss it is debited. For each grower that sells short, there may be a food manufacturer or processor that takes the opposite position, thus ensuring the hedge-trading mechanics work out.

Second, futures contracts are highly liquid, enabling all parties to exit the trade when they find it appropriate. Third, the cost of this form of price risk "insurance" is rather minimal, 0.5% to 1.0% or less of the value of the contracts. Finally, it takes the advice of a commodities broker to become proficient at this form of minimizing business risk for the grower or processor. In other words, always seek professional help in establishing a commodities hedging program for your agribusiness.

■ Questions and Problems

7-1. Given the dominating set shown in figure 7.2:

   a. Explain why the investment at point A dominates the investment at point B.

   b. Explain why the investment at point C dominates the investment at point B.

   c. Describe the relationship between the investments at points B and D. Does one dominate the other? Why or why not?

   d. Describe the relationship between the investments at points A and D. Does one dominate the other? Why or why not?

7-2. A small soybean grower expects to harvest one hundred thousand bushels in four months. The current cash price for the crop is $4.25 per

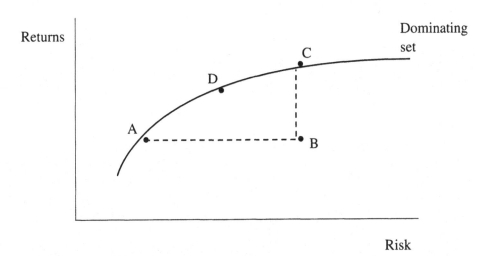

Figure 7.2. A second dominating set.

bushel. The soybean futures contract that matures in four months is currently trading at 425, or $4.25 per bushel. Each contract is for five thousand bushels.

a. How many futures contracts should the grower trade to hedge the crop price risk?

b. Which should the grower do—sell short or go long?

c. In four months the price of soybeans is $5.00 per bushel due to drought, and the futures contract is trading at 500. How much is the gain or loss on the cash crop? On the futures contracts?

d. Is this an effective hedge? Explain.

7-3. General Foods buys wheat on the cash market for its many cereal products. In January, the cash price is $2.50 per bushel. The April futures contract is trading at 250, or $2.50 per bushel.

a. To hedge against price risk, does the firm go long or sell short wheat futures?

b. In April, the cash price is $3.00 per bushel, while the April futures contract is trading at 300. What is General Foods' loss per bushel in the cash market? Its gain in the futures market?

c. Is this an effective hedge? Explain.

■ Notes

1. Steven Blank and Jeffrey McDonald, *Crop Insurance as a Risk Management Tool in California: The Untapped Market,* Research Report for the Federal Crop Insurance Corporation, September 1993.
2. "Security Blanket," *California Farmer,* September 1998.
3. *Wall Street Journal,* July 19, 1999, C16.

8

The Agricultural Lending Industry: Commercial Banks and the Farm Credit System

Agribusiness credit finance is provided by three main sources: bank lending, the Farm Credit System, and private lenders such as family members and business acquaintances. Although the majority of financing is in the form of equity or owners' contributions, a knowledge of credit sources and their requirements is key to the successful operation and growth of the agribusiness.

■ Commercial Banks

Throughout the United States today there exist approximately eight thousand independently chartered banks. We are the only country in the world where more than several hundred banks compete for business. This unique economic environment stems from our unusual banking history, beginning with the chartering of the First Bank of the United States in 1791.

Regulatory History

Following the colonies' successful independence from Great Britain, domestic banking politics involved two factions, the Federalists and the Populists. Federalists were followers of Alexander Hamilton and sought a strong central government including the regulation of banking. The First Bank of the United States, chartered in 1791, was founded to keep control of bank chartering at the federal level. In contrast, Populists distrusted the eastern seaboard money centers, and they pushed Congress to let the First Bank's charter lapse in 1811. But due to our second war with Great Britain in 1812, Congress felt the need for bank financing and chartered the Second Bank of the United States four years later. By 1832, under the veto power of Populist president Andrew Jackson, the charter was allowed to lapse.

It wasn't until 1863, with the United States embroiled in the Civil War, that Congress once again attempted to regulate banking at the federal level. During the intervening time, states established departments of banking and issued state charters. But the National Banking Act of 1863 created access to national chartering by designating the Office of the Comptroller of the Currency (OCC) as the charter agency. Since all states already had their own state chartering agencies, this was the beginning of our dual banking system.

Dual chartering provides a choice for bank founders. Either the State Department of Banking or the Office of the Comptroller of the Currency issues the charter. The majority of commercial banks are state chartered, but the larger national banks tend to be OCC chartered. Key differences lie in the national bank's mandatory participation in the Federal Reserve System and the Federal Deposit Insurance Corporation (FDIC) for deposit insurance.

Following the Civil War, banking underwent little change until 1913, when the Federal Reserve Act established our current national bank, the Federal Reserve, or Fed. Initially, the Fed's task was to provide liquidity to banks as a lender of last resort through discount loans. These short-term loans provide liquidity to banks when they run short of required deposits or reserves. Within twenty years it became obvious discount loans alone would not be enough to stave off the Great Depression.

The Fed was established as a system of twelve district banks, covering the continental United States. The oversight for its activities was placed in the hands of the Board of Governors, located in Washington, D.C. Following the stock market crash of October 1929, the United States plunged into a severe recession. Had the Fed known as much then as it does today about managing the nation's economy, this recession may have not deepened into the Great Depression. But with bank failures brought on by bad loans on securities transactions and a slumping agricultural economy, the nation's money supply shrank drastically from 1929 until 1933, at the depth of the depression. This caused a general price deflation, and surviving businesses were reluctant to expand or hire additional employees due to price and demand uncertainties. Though newly elected president Franklin Roosevelt pushed many pieces of New Deal legislation through a receptive Congress in 1933 and 1934, none of these acts made a significant dent on the nation's economy. Instead, the depression ended rather dramatically with the onset of World War II and the demand for goods, services, and employees it generated.

One significant banking act of 1933 was the Glass-Steagall Act, which separated banking from securities transactions, particularly stock issuance and margin loans, or loans by banks to stock speculators, using the stock as collateral. (Today speculators must borrow from a securities firm if they desire a margin loan.) Glass-Steagall effectively separated banking from *investment banking,* the securities business. Although this separation has been eroding in recent years due to Federal Reserve decisions, the separation of the two industries differentiates the U.S. banking system from foreign countries' systems. Most of our trading partners, such as Great Britain, Germany, and Japan, allow their banks to handle securities transactions. Because of this unfair advantage when competing internationally, the Fed had modified its regulations to allow 25% of banking revenue to come from securities activities through allied or subsidiary companies. Then, in the fall of 1999, Congress repealed the remainder of Glass-Steagall.

Following the Great Depression, Congress enacted legislation to force bank **holding companies** to comply with strict interstate branching regulations that had been in place since the 1920s. The 1970 Douglas Amendment to the 1956 Bank Holding Company Act required bank holding companies to follow states' banking laws with regard to interstate branching. The concern driving these pieces of legislation had been that big banks could drive smaller, local banks out of business, a throwback to the Populist view of the 1800s. Lost in the logic was the benefit to the consumer of additional competition. In recent years, the restrictions on interstate branching have been overturned with the passage of the Riegle-Neal Act of 1994, which allows interstate branching for all states but Texas. The result could be better service and lower costs to consumers. The jury is still out on the impact of interstate branching on bank customers.

Mergers

But while Congress's view on branching has relaxed, so has the regulators' view on mergers. The United States Justice Department, the Federal Reserve, the FDIC, and the OCC all examine proposed mergers and all seem in recent years to favor consolidation in the industry and declining numbers of bank charters. Such combinations as Wells Fargo and First Interstate, Bank of

America and Security Pacific, Washington Mutual and American Savings, and Bank of America and Nations Bank are creating much larger banks with nation-wide potential. These mergers are stock-price driven. The acquiring bank sees an opportunity to take over another bank, with the outcome to be lower costs, more territory, more customers, a wider variety of products, and, most important, satisfied shareholders (stockholders). It is likely this merger activity will continue until one or more of the regulatory agencies perceive them to be detrimental to consumers, or Congress reacts on consumers' behalf. Ultimately, the United States may have two thousand or so chartered banks, one-sixth of the number of two decades ago. The unresolved question is whether or not fewer banks will improve the lot of the consumer.

Bank Balance Sheet Basics

As financial intermediaries, banks accept deposits, make investments, and hope to profit on the interest rate differential or spread. A basic balance sheet is provided in table 8.1 below.

Table 8.1. Basic bank balance sheet ($ $\times$ 1,000)

Assets		Liabilities & Equity	
Cash & reserves	$ 2,000	Demand deposits	$ 20,000
Securities	26,000	Savings deposits	65,000
Consumer loans	20,000	Bank debentures	17,000
Real estate loans	25,000	Other borrowings	0
Commercial loans	31,000	Common stock @ par	2,000
Less loan loss reserves	2,000	Excess paid in capital	1,000
Premises & equipment	4,000	Retained surplus	5,000
Total assets	$110,000	Total liabilities & equity	$110,000

Liabilities

Traditionally, banks obtained the majority of their deposits from demand deposits, the consumers' checking accounts. The bank earns maintenance fees on these accounts and may provide overdraft protection for additional fee income. This mainstay source of bank funds has been eroding as a percentage of total deposits since the 1970s, when investment banking firms introduced uninsured money market checking accounts, high-yield demand deposit accounts lacking the FDIC deposit insurance. Since that decade, banks have had to compete for deposits by providing additional services and paying interest to depositors. Today, demand deposits represent several types of checking accounts, including the traditional non-interest-bearing checking, interest-bearing checking, and bank money market checking accounts.

Savings deposits include the classic passbook savings, small denomination certificates of deposit, larger negotiable certificates of deposits (CDs), and special accounts such as college education savings funds, individual retirement accounts (IRAs), or other long-term deposits. Because savings deposits are not technically available upon the demand of the depositor, the banking industry provides a higher interest rate than for checking deposits. Certificates of deposit usually incur a penalty for early withdrawal.

Bank debentures are bond instruments issued by the bank or its holding company. Bonds pay periodic interest to the investor, along with a lump sum of principal at maturity. Although not a major source of funds for banks, the long-term nature of the obligation makes this a good form of borrowing when interest rates are favorable. Other borrowings include the above noted discount loans from the Fed, overnight loans from other banks known as Fed Funds, and less-common repurchase agreements or repos, short-term borrowings from cash-flush corporations to meet the bank's temporary liquidity needs. For smaller-sized balance sheet banks, the use of other borrowings is minimal.

Equity

The remaining accounts on the right side of the balance sheet are ownership accounts, or equity in the bank. The common stock at par account is the value of shares of stock sold based on a stated par value, frequently ten dollars a share. Excess paid-in capital is the additional dollars raised by the bank (or its holding company) when the shares are first sold, if they sell for more than the designated par value per share. Suppose a bank sells two hundred thousand shares of ten-dollar par value common stock to raise initial capital. If the shares each sell for fifteen dollars a share instead of ten dollars, then the excess would be five dollars a share, or a total of one million dollars excess paid-in capital. For most bank balance sheets, there is some value in this account, as it is unlikely the stock will sell for exactly its par value.

Retained surplus, which is also called undivided profits and retained earnings, is the net income earned by the bank in previous periods that has not been paid out to stockholders as dividend income. When an incorporated business makes a profit, this is represented by an increase in assets (left side of balance sheet), and an increase in the balance sheet account on the right side of the balance sheet, retained earnings. There has been an historical trend to label this retained profit as retained surplus or undivided profits. Regardless of the label, the figure represents earlier net income that remained with the bank; that is, it was retained for use in the business. The balance sheet remains in balance because the increase on the asset side is matched by a larger balance in retained surplus. How the bank utilizes prior period profits is a matter of asset financial management.

Assets

Cash and reserves are funds kept either at the bank's retail location in its vault or in teller cash drawers, or at its district Federal Reserve Bank branch. To control monetary policy and strengthen the soundness of our national banking system, the Fed requires approximately 10% of bank demand deposits to be placed on reserve, in the vault as cash or deposited at the Fed's branch. This reserve requirement frequently causes banks to scramble during business hours to obtain reserves from other banks in the form of the liability Fed Funds, noted above. In effect, the Fed will place operating restrictions on the bank if it discovers insufficient reserves. Contrary to one's intuition, cash and reserves are not really sources of liquidity for a bank; they merely meet the day-to-day cash needs of small customers.

A key source of liquidity for the bank is its securities portfolio. To minimize risk, banks prefer holding government securities such as Treasury Bills, which yield interest revenue greater than the cost of deposit interest under normal credit market conditions. But securities do not yield sufficient interest to make the bank highly profitable. The main source of revenue and hence net income for banks is its loan portfolio.

Consumer loans include credit card debt, short-term installment debt such as automobile financing and other consumer products, and specialized loans such as student financial aid. These loans are short term or revolving in nature, pay a fairly high interest rate, and comprise a large part of most banks' loan portfolios. Revolving debt must be paid off periodically, while most credit cards do not require a zero balance; instead, the borrower must make the minimum payment. From the bank's perspective, a large credit card balance on the account of a responsible customer is an ideal loan, paying a high rate of interest with a short term. Consumers should use this source of financing wisely, making prompt payment of the balance due to avoid paying the expensive interest costs.

Real-estate loans can be both home loans and other categories, including construction financing, multifamily housing loans, raw land financing for development, and commercial property financing. Many banks view construction financing as key to their operations, earning high effective rates of interest due to the points and other fees paid by the borrower, and putting the bank in the position of marketing permanent financing to the ultimate investor. Other real-estate lending, including agricultural properties, involves the underwriting techniques and ratio analysis discussed in chapter 6. Real-estate loans, with the exception of construction lending, tend to be long term. This implies higher interest rates under normal yield curve conditions, but also greater interest rate risk. If a bank provides a twenty-five-year farm loan with an 8% interest rate when long-term interest rates are near 8%, it is using its funds satisfactorily. But if market interest rates rise to 10%, then this 8% is now below market. The loan can only be sold at a loss. So real-estate lending creates interest rate risk for the bank, which can be mitigated in several ways.

Commercial loans include revolving lines of credit, short-term inventory financing, medium-term equipment financing, and other business borrowings. We will address agribusiness commercial loans in a following section. The key bank issues for commercial lending include the difficulty of underwriting the application, the relatively short-term nature of these loans, the goodwill (or possible *bad* will) they can generate, and their relatively large balance sheet and revenue contributions. Commercial loans are important from both an income and customer relations standpoint. Some banks have created new customer development departments or positions, primarily to enhance the bank's profile in the commercial loan market. But due to the relatively high-risk nature of these loans, interest rates are lucrative. Also, business borrowers are assumed by regulators not to need the consumer protection that private individuals receive, making these loans less structured for the lender. For example, there is no Regulation Z APR requirement for an inventory loan to an agribusiness supplier.

Loan loss reserves is an account known as a "contra-asset," that is, it counters or lowers the value of its related asset accounts. Banks have for years experienced losses on their loan portfolios. With experience, the chief credit officer or asset committee can estimate the proportion of loans that are likely to go into default and end up uncollectible. The loan loss reserve account stores the value of current and former estimates by the bank of future uncollectible loans. Although the specific uncollectible loans cannot be predicted (if they could, why make them?), experience can help the bank make a reasonable estimate. Since some small proportion of the loans will inevitably go into default, this account reduces the total value of the loans in the portfolio and the value of total assets.

Bank premises and equipment tend to have a smaller account value on the asset side of the balance sheet. Banks tend to lease retail space and invest in human capital rather than heavy equipment. The computer systems and actual transaction processing equipment are included in this account. If the bank owns its corporate headquarters or home office real estate, that would be included. Banks view items in this account as non-earning assets; that is, there is no interest income to be derived from investment in premises and equipment. Therefore, the value of this account category is kept to a minimum.

As we have seen, the balance sheet displays the bank's sources and uses of funds. Since its objective is to make a profit on its investments (assets) relative to its sources of funds (liabilities and equity), management strives to keep asset yields high while minimizing the interest cost of its liabilities and equity, known as its cost of funds.

Bank Agribusiness Loans

Banks of all different sizes are involved in agribusiness finance. The larger, urban or money center banks tend to finance large-scale poultry and livestock operations and specialty crops. Regional or statewide banks have seen their share of ag loans increase while rural banks with relatively small total assets have experienced a declining market share of ag loans. Banks with more than 40% of their assets tied up in ag loans tend to be smaller, rural institutions.[1]

Some banks label their ag loans separately on the balance sheet. More commonly, agricultural loans are categorized similarly to our balance sheet in table 8.1. One lender informally classifies its agribusiness loans as either lines of credit, crop-production loans, equipment loans, or farm and ranch loans.[2]

Agribusiness suppliers need inventory financing; this is usually a line of credit. For equipment, medium-term fully or partially amortizing loans are common, while the farm and ranch financing is long-term fully amortizing. As noted in chapter 6, several key ratios assist the commercial bank in evaluating and underwriting agribusiness loans. Ratios recommended by the Farm Financial Standards Council include term debt and capital lease coverage, and the capital replacement and term debt repayment margin (see repayment capacity ratios in chapter 6). As previously indicated, California ag lenders prefer the current, debt-to-equity, term debt and capital lease coverage, and loan-to-value ratios. The current ratio is useful for lines of credit, crop production, and inventory loan analysis, while the other three are key to underwriting longer-term loans for real estate and equipment.

Loan Underwriting Example

All American Farms, our chapters 5 and 6 example, has requested a new farm real-estate loan to expand its operation. The owners have located a one-eighth-section contiguous parcel for sale at $1,800 an acre. Financing will be provided by Midwest Bank, a small regional bank in Iowa. The terms of the loan include 35% down, the balance financed over twenty-five years at 9%, fully amortizing, with monthly payments. To underwrite the loan, Midwest Bank's loan officer has compiled the following summary financial data for 1998:

Land purchase price	$144,000.00
Down payment (35%)	50,400.00
Additional crop sales	80,000.00
Additional op. farm expenses	54,000.00
Monthly payment	785.49
First-year interest expense	8,382.00

To calculate the three long-term ratios, first we note that the loan-to-value ratio is a given. The lender requires a minimum of 35% down and this generates a loan to value of 65% (the complement of the down payment percentage). However, the loan officer must obtain an appraisal that confirms the value of $144,000. Assuming the appraisal is satisfactory, then the loan to value is 65% and minimally meets Midwest Bank's requirement.

Next, the new forecasted data above are combined with the financial statement data for All American Farms provided in chapters 5 and 6 to determine the debt-to-equity and coverage ratios. Using 1998 year-end balance sheet figures from table 5.3, the projected debt to equity becomes:

$$\frac{\text{Total debt}}{\text{Owner's equity}} = \frac{246,000 + 93,600}{559,000} = 60.75\%$$

which is satisfactory under the industry and Midwest Bank guideline of less than 100%. The projected term debt and capital lease coverage ratio would be calculated as follows:

$$\frac{\text{Coverage}}{\text{Ratio}} =$$

$$\frac{94,000 + 26,000 + 1,000 + 20,000 + 52,000 + 23,382 - 22,000 - 40,000}{21,000 + 9,426}$$

$$= 5.07{:}1$$

which is well above the industry and Midwest Bank standard of 1.25:1.

Based on the ratio analysis, appraisal, and credit history of the owners of All American Farms, it is likely the loan officer would recommend approval of this farm real-estate loan.

■ The Farm Credit System

The Farm Credit System (FCS) is a source of loanable funds that is unique to the U.S. agricultural sector. FCS loans are made only to agricultural and aquatic producers, agribusinesses that provide on-farm services, agricultural cooperatives, and rural residents needing residential financing. The system is organized as a series of cooperatives, called associations, in which the borrowers are the owner/members. Borrowers elect, from their own ranks, the members of the boards of directors of the local associations. As of 1997, FCS total farm-business debt amounted to $42.3 billion, which was second only to the agricultural loan market share of commercial banks at $67 billion.[3]

History

The Farm Credit System originated during an era when agricultural cooperatives of all types were being widely promoted and established across the

United States. The Federal Farm Loan Act of 1916 created the original twelve district Federal Land Banks to which the local Federal Land Bank Associations (FLBAs) reported. The Federal Land Banks were authorized to make long-term real-estate loans to farmers and ranchers.

In 1923, the Federal Intermediate Credit Banks and, in 1933, the Production Credit Associations (PCAs) were established by acts of Congress. These FCS institutions were created to provide short-term and intermediate-term non-real-estate loans to agricultural producers. Also in 1933, the Banks for Cooperatives (BCs) were set up to lend funds to agricultural cooperatives involved in the marketing of farm products and in the furnishing of supplies and services to farmers.

Although the initial capital needed to start all FCS institutions came from the federal government, the government has long since been paid back. In 1947, the FLBAs became completely owned by the association members, and the same thing occurred for the PCAs and BCs in 1968.[4] The FCS obtains the money that it lends to qualified borrowers from the public sale of systemwide securities (bonds and notes). The FCS is not a depository institution. It does not depend on deposits in the form of checking accounts, savings accounts, or certificates of deposit like a commercial bank as a source of funds. FCS securities are categorized as government agency securities, but they do not carry any guarantee of repayment by the federal government. The fiscal agent that administers the issuance and repayment of FCS bonds is the Federal Farm Credit Banks Funding Corporation, which has offices in Jersey City, New Jersey, near the New York City financial district.

Regulatory power over the Farm Credit System has been the responsibility of the Farm Credit Administration (FCA) since 1933. The headquarters of the FCA are located in McLean, Virginia. The FCA is administered by a three-member board, whose members are appointed by the president of the United States. Staff members of the FCA are responsible for auditing the system associations to insure that loans and loan portfolios remain strong.

Until the 1980s, the Farm Credit System continued to expand its loan volume and increase the number of its lending institutions. In 1983, there were 895 separate FCS associations lending to farmers and ranchers across the country, 474 FLBAs and 421 PCAs.[5] These local associations were divided into twelve districts each with its own Farm Credit Bank. Farm business loan volume held by the FCS peaked at $64.7 billion in 1984. This is well ahead of the $47.2 billion market share of commercial banks at that time.[6]

Unfortunately, the farm financial crisis of the 1980s had devastating effects on the performance and financial stability of the FCS, and ultimately a massive reorganization of the system was required. The economic distress of farmers and ranchers during the early and mid-1980s often resulted in delinquency and default on their loan payments and consequently in economic distress for their lenders. The FCS specializes in debt financing for agribusinesses, so it did not have much leeway to diversify its loan portfolio. Since the systemwide bonds were the joint liabilities of all FCS associations, the associations that were surviving the tough times were expected to help support the associations that were in financial trouble. Needless to say, interdistrict and interassociation relations and morale were not at an optimum level throughout these crisis times.

Congress stepped in and offered assistance in the form of the Agricultural Credit Act of 1987. A government-backed loan package of up to $4 billion was established to support the FCS, and reorganization of the FCS was mandated. The changes stipulated in the Agricultural Credit Act of 1987 have resulted in the much more streamlined system that has evolved, and is still evolving, in the present.

Current Structure and Practices

Reorganization through mergers and consolidations has vastly reduced the number of administrative organizations and associations operating in the system. There are now six district Farm Credit Banks instead of twelve. The number of Banks for Cooperatives has been reduced from thirteen to two, one located in Denver, Colorado, and another in St. Paul, Minnesota. Instead of the 421 PCAs of 1983, there were 64 PCAs in 1998. The 474 separate FLBAs that existed in 1983 became 40 FLBAs and 32 FLCAs in 1998. An FLCA is a Federal Land Credit Association that, like an FLBA, handles long-term real-estate loans. The difference is that an FLCA directly grants and funds its own loans with monies it obtains from a Farm Credit Bank while an FLBA takes loan applications and services loans, with the loan-granting decision and loan funding the responsibilities of the Farm Credit Bank. Additionally, the establishment of Agricultural Credit Associations (ACAs) has been encouraged. In 1998, there were 57 ACAs across the United States. An ACA can directly grant both long-term loans like an FLCA and short- and intermediate-term loans like a PCA.[7] Figure 8.1 shows the regional authorities of the six Farm Credit Banks and one Agricultural Credit Bank, which was created by a merger of a Farm Credit Bank and a Bank for Cooperatives.

During the late 1980s and the 1990s, the Farm Credit System increased its net earnings from year to year. Systemwide net income was $1.27 billion in 1997[8] compared to a net loss of $2.7 billion in 1985.[9] Probable reasons for this turnaround are the general improvement in the agricultural economy, lower interest rates, more conservative loan underwriting policies, and increased operating efficiencies from the reorganization of the FCS. Of the $4 billion federal "bailout" specified in the Agricultural Credit Act of 1987, less than $1.3 billion was used, and the FCS has been paying back the funds that were borrowed ahead of schedule.[10]

Figure 8.1. Farm Credit System Banks Chartered Territories
(As of July 1, 1999)

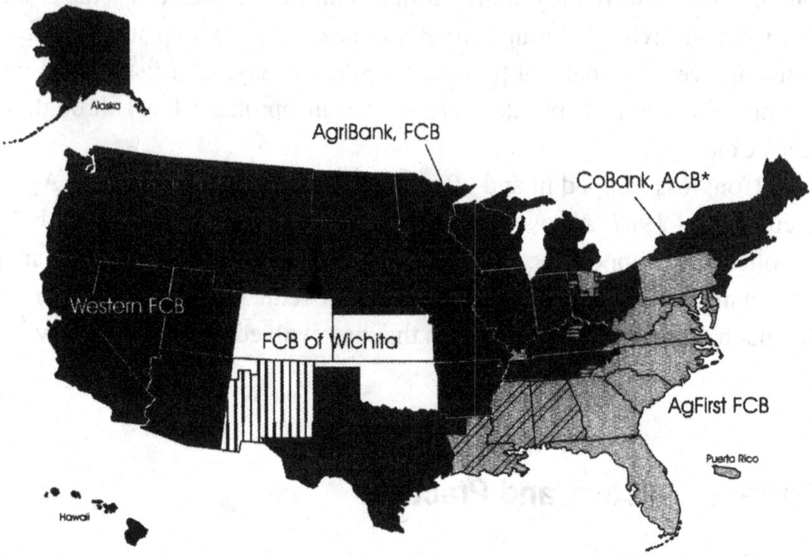

AgAmerica, FCB	Western FCB	FCB of Wichita	FCB of Texas	AgriBank, FCB	CoBank, ACB	AgFirst FCB
1 ACA	5 ACAs	13 FLBAs	19 FLBAs	8 ACAs	4 ACAs	34 ACAs
1 FLCA	11 FLCAs	9 FLCAs	1 FLCA	18 FLCAs		1 PCA
1 PCA	10 PCAs	18 PCAs	15 PCAs	18 PCAs		

The PCA of New Mexico and PCA of Eastern New Mexico are funded by the FCB of Texas. The PCA of Southern New Mexico is funded by the FCB of Wichita.

The FLBAs in Alabama, Louisiana, and Mississippi generate and service loans for the FCB of Texas. The First South PCA is funded by AgFirst FCB.

The Northwest Louisiana PCA is funded by the FCB of Texas.

*The CoBank, ACB is headquartered in Denver, Colorado, and serves cooperatives nationwide and ACAs in the indicated area.

The AG CREDIT, ACA (Ohio), Central Kentucky ACA (Kentucky), Chattanooga ACA (Tennessee), and Jackson Purchase ACA (Kentucky) are funded by the AgFirst FCB.

The Mid-America ACA, funded by AgriBank, FCB, is also authorized to lend in this territory.

The Eastern Idaho ACA is funded by the Western FCB.

Source: Farm Credit Administration, McLean, VA

The typical agricultural borrower may not even have noticed the recent FCS merger activity if his or her branch office was not closed. A borrower still deals with a loan officer located in a branch office. That loan officer can handle long-term loans if he or she works for an FLBA, FLCA, or ACA, and can handle short-term (such as a line of credit) or intermediate-term loans if he or she works for a PCA or an ACA. The loan-approval process is similar to what it would be for a commercial bank or any other institutional lender. The FCS loan officer usually requires a current balance sheet, the most recent income statement, historical income statements, a cash-flow budget, and three years of tax returns. Often the loan officer will make a physical inspection of the borrower's place of business. FCS loan officers are frequently experts on the production and marketing of the commodities raised within their areas. If an appraisal is needed, an appraiser who works for the local association will be sent out to complete the appraisal as soon as possible. If an association does not have its own appraiser on staff, then a fee appraiser may be hired. Whichever the case, the loan officer and the appraiser should be two separate people in order to avoid any potential conflict of interest. The loan analysis function and underwriting decision rests with the loan officer and the association superiors for PCAs, FLCAs, and ACAs. In the case of FLBAs, all the information that has been gathered and analyzed is sent along to the Farm Credit Bank for a final decision.

Long-term loans are usually used to finance the purchase of real estate, the development of orchards or vineyards, or the construction of farm buildings and other improvements. Also, the borrower may wish to refinance an existing loan and negotiate a new interest rate, principal, or term. The FCS offers long-term loans with terms between five and forty years and with fixed or variable interest rates. Most loans are made for terms between twenty and thirty years. Sizable down payments of at least 30% to 35% of the appraised value of the real estate are normally required. Amortized payments are made by the borrower, usually one per year. The real estate in question is pledged as security for the note, and the FCS will not grant a real-estate loan unless it is in first position on the collateral.

Short-term loans, that is, loans with terms of less than one year, are usually set up to pay for operating costs. Due to the seasonality of agricultural production, many types of farms and ranches require large expenditures for operating costs during the growing season and only realize cash revenues at one time during the year. Consequently, lines of credit are a popular product of the FCS, and a producer's line of credit will be set up based on his or her cash-flow budget. Interest is only charged on the outstanding principal for the number of days it remains unpaid. The primary collateral for short-term loans is the crop or livestock being raised, but other assets such as equipment or real estate can also be requested as secondary collateral.

Intermediate-term loans commonly finance the purchase of equipment or breeding stock. The terms run between one and ten years, three to five years being the most common. In many cases the property being financed is the total collateral that is required. With the increased popularity of capital leases, the FCS set up the Farm Credit Leasing Services Corporation in 1983. This organization is owned by the Farm Credit Banks and is headquartered in

Minneapolis, but many of the PCA and ACA offices in the field will process equipment leases for their customers. Tractors, trucks and other vehicles, combines, farm implements, dairy equipment, and processing equipment are well suited to leasing contracts. Lease placements of the Farm Credit Leasing Services Corporation were $480 million during its 1997 fiscal year, and it had 41,000 leases with 8,700 customers as of September 1997.[11]

As previously stated, FCS institutions are cooperatives, and the borrowers are the owners. This ownership is achieved through association stock that is issued when funds are borrowed from an FCS institution. For many associations the stock is equal to a given percentage, 1% for example, of the principal requested by the borrower. For other associations, the stock is a fixed fee such as one thousand dollars. In some situations, the stock becomes part of the unpaid principal, and the borrower pays interest on the stock as well as on the original principal advanced. In other cases, no interest will be charged on the stock. As the borrower pays back principal and interest, he or she does not have to pay back the stock; the stock will be wiped out on the books of the association when the loan is paid off. In recent years, FCS associations have been able to lower the stock percentage and fixed-fee requirements.

The systemwide debt securities sold to investors and used to finance the lending activities of the FCS are of three types: bonds, discount notes, and global debt securities. The bonds have maturities of three, six, and twelve months and are scheduled to be sold on a monthly basis. However, when additional funds are needed, unscheduled bonds can be sold by the Federal Farm Credit Banks Funding Corporation. These bonds can be callable or uncallable. A callable bond contains a call provision that allows the issuer to redeem the bond before its maturity date by paying the investor the principal and all previously unpaid interest as of the call date. An uncallable bond cannot be redeemed until its maturity date. The discount notes of the FCS can be issued on a daily basis and have maturities of 1 to 365 days. The global debt securities are issued in dollars and in several foreign currencies. They can have maturities ranging from 30 days to 30 years. The Farm Credit Service Insurance Corporation was set up in 1989 to insure that principal and interest are paid in a timely manner on all three types of FCS debt securities.

■ Questions and Problems

8-1.   Explain the "dual chartering" system for U.S. banking.

8-2.   A new bank sells 500,000 shares of its ten-dollar par value stock for fifteen dollars per share. Following the sale, what are the values for its *common stock @ par* and *excess paid in capital accounts*?

8-3. A simplified bank balance sheet for 1999 appears below.

Assets		Liabilities and Equity	
Cash and reserves	?	Demand deposits	$10,000
Securities	24,000	Other borrowings	71,000
Loan portfolio	62,000	Paid in capital	4,000
Other assets	?	Retained surplus	5,000
Total assets	$90,000	Total liabilities and equity	$90,000

Cash and reserves are 10% of demand deposits and other assets make up the difference of total assets.

   a. Complete the above balance sheet.
   b. In the year 2000, the bank receives an additional $10,000 in deposits and increases its retained surplus by $5,000. Cash and reserves maintain their 10% ratio to demand deposits, while any additional increase in assets is absorbed by the loan portfolio. There are no other changes to the balance sheet. Prepare the year 2000 ending balance sheet.

8-4. At which types of Farm Credit System offices (PCA, FLBA, FLCA, ACA, or BC) would the following loans be available? There may be more than one answer for each question.
   a. a loan for the purchase of a farm
   b. a line of credit for a greenhouse to pay for operating expenses for the upcoming year
   c. a loan for a farm cooperative to expand its storage facilities
   d. an operating loan for an agricultural chemical company that is heavily involved in the on-farm application of its chemicals
   e. a real-estate refinancing loan to a dairyman to build a new milking barn
   f. a loan for a farmer to purchase a tractor

■ Notes

1. Peter J. Barry, Paul N. Ellinger, John A. Hopkin, and C. B. Baker, *Financial Management in Agriculture,* 5th ed. (Interstate Press, 1995).
2. Christa Kodl (loan officer), Mid State Bank, Telephone interview, September 8, 1998.
3. U.S. Department of Agriculture, Economic Research Service, *Agricultural Income and Finance Situation and Outlook Report,* February 1999, 54.
4. Ben Sunbury, *The Fall of the Farm Credit Empire* (Iowa State University Press, 1990), 12.
5. Warren F. Lee and George D. Irwin, "Restructuring the Farm Credit System: A Progress Report," *Agricultural Finance Review* 56 (1996): 12.
6. USDA, *Agricultural Income,* 1999, 54.
7. Farm Credit Administration, Mid-Year Report (Farm Credit Administration, 1998), 56.
8. USDA, *Agricultural Income,* 1999, 19.
9. U.S. Department of Agriculture, Economic Research Service, *Agricultural Income and Finance Situation and Outlook Report,* February 1992, 19.
10. Lee and Irwin, "Restructuring the Farm Credit System," 13.
11. Farm Credit Administration, *Mid-Year Report* (Farm Credit Administration, 1997), 19.

9
Other Agribusiness Lenders

Vendor/Trade Credit
Life Insurance Financing
Farm Service Agency
Federal Agricultural Mortgage Corporation (Farmer Mac) and the
    Secondary Market
Private Financing
Beyond the Year Two Thousand
Questions and Problems

In the previous chapter, we learned the role of commercial banks and the Farm Credit System in financing agribusiness. Now we turn to several other sources, and include an explanation of the burgeoning secondary market for ag real-estate mortgages.

■ Vendor/Trade Credit

A **vendor** is a merchant or businessperson selling products to businesses and the public. **Trade credit** is financing provided by a vendor. Whether the commodity grower needs farm inputs from a farm-supply merchant, a vintner is buying a truck from a dealer for bottled wine delivery, or an ag product manufacturer, such as John Deere or International Harvester, is acquiring parts from another manufacturer for use in its equipment manufacturing, if the financing is provided by the seller it is deemed trade credit. Trade credit can be for farm inputs such as seed, feed, fuel, and fertilizer, or for manufacturing related purchases, but does not include real-estate financing. Because of the seasonal nature of agriculture, most input purchases could not take place without some form of short-term financing. Trade credit and other working capital loans provide the means for producers to obtain the raw materials they need, and make payment when the commodity has finally been harvested and sold. Undoubtedly, trade credit increases sales for farm supply vendors.

Most trade credit consists of an open account, such as consumers can hold with retail stores. This requires a billing cycle, the responsibility of the vendor, who requests payment from the purchaser at least monthly. **Credit terms** are the contractual agreements between the parties as to the timing of payment, any discount for early payment, and penalties for late payment or nonpayment. Credit terms are usually described in a "shorthand" terminology in which the possible discount for early payment is quoted as a percentage, immediately followed by the payment date to receive the discount. Most terms then state the final due date for payment after which the payment is late and likely to incur a penalty as specified in the trade credit agreement. A common set of credit terms is 2/10, n30, stated as two-ten, net thirty, meaning that a 2% discount will be granted by the vendor (creditor) if payment is received within ten days of the invoice, and the entire amount of the purchase is due within thirty days of the invoice.

Some vendors do not provide early payment discounts, while others require that the entire payment be received by the end of the month of the invoice, or EOM. Whether the vendor is lenient or strict with customers is a matter of credit policy. Some agribusiness firms hire credit managers to formulate or administer the firm's credit policy. This specialized financial management position requires careful analysis of the effects of credit terms and enforcement on the firm's profitability. Too liberal credit terms, such as providing a large early payment discount, may erode profits needlessly, while an overly strict or "get-tough" credit policy may drive away potential customers and reduce sales and profits. When the effect on profits of a change in credit policy is difficult to gauge, the credit manager can use past experience and cost-benefit analysis to help in decision making. For example, if an additional 1% early payment discount is estimated to increase sales 5%, then as long as the firm has a sufficient contribution margin for its products, this change in credit policy would be justified. Advanced financial management texts analyze change in credit policy issues in detail.

For the typical agribusiness, a key concern is that of whether or not to take advantage of early payment discounts. Under the traditional credit terms of 2/10, n30, the early payment must be made by the tenth day after the invoice. If the customer forgoes the discount, the payment must be made by the thirtieth day. So, by forgoing the discount, the customer receives an additional twenty days financing. What is the approximate interest rate on this financing? Assume, for example, that the customer purchases $1,000 per day of farm inputs, 360 days per year. This totals annual purchases of $360,000. If the customer takes the early payment discount, he or she saves 2% of $360,000, or $7,200 per year. Since the customer will always pay each invoice on the tenth day, there will be $10,000 owed in accounts payable at all times. Now, assume instead that the customer forgoes the early payment discount to take advantage of the additional twenty days financing, paying all invoices on the thirtieth day. Now the balance in accounts payable has increased to $30,000. So the customer has obtained an additional $20,000 financing. But the interest rate is usurious! The effective interest rate, i, is found as follows:

$$i = \frac{\$7,200}{\$20,000} = 36\%$$

An informal survey of California Central Coast ag vendors indicates that the trend is toward smaller or even no early payment discounts. Typical credit terms today could be 1/10, end of month. An analysis of these terms can be performed as in the above example, with an effective interest rate of approximately 24%. At that rate, it still pays to take discounts when available. The lesson: never forgo early payment discounts to finance purchases if other financing is available.

■ Life Insurance Financing

Prior to the establishment of the Farm Credit System, life insurance companies were major providers of real-estate financing for agriculture. Because of the life insurance industry's increased investment in corporate bonds and other securities, ag real-estate loans have declined to less than 0.5% of total assets in the industry. Still, this amounts to $11.3 billion nationally, making life insurance firms a major source of ag mortgage financing.[1]

Life insurers tend to provide larger loan amounts, with the average farm mortgage outstanding being approximately $730,000. This amount understates the size of new loans as many of the outstanding loans are well "seasoned," meaning the borrower has been making payments for some time with the balance correspondingly reduced.[2] The loan application is typically handled by a loan correspondent or mortgage banker, which can be a small, independent loan company or a local branch of a regional or national mortgage banking firm. Since mortgage bankers derive much of their revenues by providing servicing, which is the periodic billing, amortizing, and legal maintenance of the loan, some life insurance loans are essentially managed by the local loan company, while other life insurance lenders have their own servicing departments.

Several trends in life insurance ag real-estate lending are worth noting. During the past decade, the number of life insurance companies providing this financing declined to approximately twenty firms. Their total loans outstanding were 15,800 as of June 1998. Although delinquent loans (defined as one or more payments in arrears) increased in 1998 to 1.8% of total ag loans, foreclosures are at two-decade lows, well below the rates for nonagricultural mortgages.[3] And though the proportion of life insurance holdings of total agricultural real-estate debt has fluctuated throughout the twentieth century, the industry now holds 11.2% of the total farm real-estate loans outstanding, not far from the 12% it held around the turn of the century.[4] While the total dollar value of ag real-estate loans in life insurance companies' portfolios has continued to increase, the industry's share of total ag loans peaked at 25.1% in 1956. Borrower demographics have also shifted somewhat, away from the traditional farm mortgagor, and toward ag suppliers, timber-industry firms, and specialty firms. This is consistent with the life insurance industry's interest in increasing its commercial property mortgage holdings. Finally, the insurance industry has steadily increased its share of ownership of ag properties, now topping $3.2 billion in direct farmland investments.[5]

Ag mortgages provided by the industry come in a variety of amortization programs. Traditional, fixed interest rate, fully amortizing loans remain popular, while the fixed-rate loan with balloon payment is a common alternative, requiring the borrower to refinance or provide cash to pay off the balloon payment when due. Some life insurance ag loans are adjustable rate, based on local or national indexes, much like many commercial property real-estate mortgages and the popular ARM residential mortgage. It seems likely the life insurance industry will continue to be a major provider of real-estate financing to agribusiness, while the share of ag loans relative to the industry's total assets will hold steady or even decline from its current level of less than 0.5%.

■ Farm Service Agency

The Farm Service Agency (FSA) was founded in 1994 following a reorganization of the U.S. Department of Agriculture. The lending functions of the FSA were previously administered by the Farmers' Home Administration (FmHA), which has existed since 1937. The FSA provides lender assistance to farmers in both direct loans and guaranteed loans, which are granted by conventional lenders such as commercial banks. Total 1999 outstanding FSA farm debt is approaching sixteen billion dollars, 9% of the total U.S. farm debt.[6] FSA's loan programs are summarized in figure 9.1.

Direct Loan Programs	Guaranteed Loan Program
Farm Ownership Loans (FO)	Farm Ownership Loans (FO)
Farm Operating Loans (OL)	Farm Operating Loans (OL)
Emergency Loans	

Figure 9.1. Farm Service Agency loan programs.

FSA's mission statement underscores its commitment to new and disadvantaged farmers, along with credit counseling and monitoring, through the above programs. The Guaranteed Loan Program (GLP) provides FSA's principal guarantee to the commercial lender for up to 95% of the total balance, providing the loan meets FSA's strict collateral criteria. This program's underwriting criteria are more stringent than the Direct Loan Program (DLP), which provides backup financing for those potential borrowers who fail to meet the GLP guidelines. Both the GLP and the DLP loans provide financing for both property ownership and working capital or operational funds. Operating loans have terms up to seven years, while FO loans provide amortization over thirty years. Loan limits range from $200,000 for Direct Operating Loans up to $500,000 for Direct Emergency Loans, to a maximum of $700,000 for Guaranteed Farm Ownership and Operating Loans.

FSA programs differ from other lenders in this chapter in that their focus is on the farmer, not the supporting agribusinesses. They work closely with farmers through their network of offices throughout the United States. The trend of the past two decades has been for funding from FSA, and the FmHA before, to decline. Total USDA-related farm debt in 1986 was over $29 billion, while today FSA's total stands at $15.9 billion.[7] For the past several years funds provided by Congress have been depleted in the DLP while there has remained excess Guaranteed Operating Loan funds at the end of the fiscal year. Emergency Loan volume declined in 1999, but this is dependent on weather and other climatological events. With FSA's main emphasis on the beginning farmer, loan originations should continue trending downward in the foreseeable future.

■ Federal Agricultural Mortgage Corporation (Farmer Mac) and the Secondary Market

Originally founded by the Farm Credit Act of 1971, authorization for Farmer Mac's secondary market, or loan purchase activities, wasn't granted until the 1987 Agricultural Credit Act. In concept, it was to provide a market for new and existing ag real-estate loans, in effect connecting ag lenders and borrowers with Wall Street. Four years following the 1987 act, it incorporated, sold its first stock offering, and transacted its initial secondary-market activities. The Farmer Mac I Program, providing its guarantee against default on the mortgages in a pool of loans, required lenders to gain approval as "poolers" of ag real-estate loans in order to sell mortgage-backed securities with Farmer Mac's guarantee. By 1995, less than one billion dollars in ag mortgage collateral for ag mortgage-backed securities had been guaranteed under Farmer Mac I. Problems had surfaced due to the relatively flat demand for ag real-estate loans this decade, the tedious pooling certification process required by Farmer Mac, and a requirement that loan poolers show a 10% contingent interest as a liability on their balance sheets. This put loan poolers under bank regulator scrutiny and discouraged use of Farmer Mac I.

Fortunately, in February 1996 Congress enacted the Farm Credit System Reform Act, vastly improving Farmer Mac's operating guidelines. The act overhauled the mortgage-backed securities issuance for Farmer Mac, provided assistance for raising capital, and eliminated the 10% contingent liability reporting requirement for lenders. Since 1996, Farmer Mac has become profitable and has begun to operate much like its larger cousins, Fannie Mae and Freddie Mac. Currently, it provides three ag mortgage-backed securities programs, Farmer Mac I, Farmer Mac II, and the Ag Vantage Program. Farmer Mac I and II are direct loan purchase programs from the lender, while Ag Vantage is a form of bond financing for ag lenders, in effect borrowing from Farmer Mac to make additional ag loans.

The Farmer Mac I program is designed to purchase ag lender real-estate loans for both working farms and rural properties. Its Full Time Farm loan purchases provide financing for working farms on purchase or refinance basis, while the Part Time Farm loan purchases finance rural properties meeting Farmer Mac's guidelines. In both branches of the program, Farmer Mac is the issuer of the Agricultural Mortgage-Backed Securities, or AMBSs. The lending institution merely makes and sells the loan, and may provide some loan servicing. Under Farmer Mac II, only USDA-guaranteed loans, such as Farm Service Agency FO or OL loans, can be sold. Farmer Mac purchases the guaranteed portion of the loan, while the lender remains invested in that portion of the loan not guaranteed by FSA, known as a participation interest. This does not appear to be the problem it once was as the lender is no longer required to disclose the participation interest as a contingent liability.

There are several loan sale contracts provided by Farmer Mac. These contracts differ as to the timing of the delivery of the loan to Farmer Mac. Lenders can sell loans at Farmer Mac's "cash window" for immediate delivery when the loan is finally closed. This implies the lender would deliver title to the loan, including any paperwork or electronic documents, in seven days. Alternatively, the lender could "forward contract" with Farmer Mac, selling a loan that is not yet closed, to deliver title to Farmer Mac in four or eight weeks in the future. This means the loan process is not yet finished and there is a chance the loan will not actually close, a loan industry problem known as fallout. A third alternative for the lender is to "swap" loans for Farmer Mac's AMBSs. This swap transaction has the effect of taking loans out of the lender's loan portfolio, and therefore off of its balance sheet as an asset, and replacing the loans on balance sheet with Farmer Mac I or II AMBSs. The AMBSs have the advantages of being more liquid and very low risk. This third method of marketing loans to

Farmer Mac has been gaining popularity the past several years and is similar to the ongoing swap programs maintained by Fannie Mae (Federal National Mortgage Association) and Freddie Mac (Federal Home Loan Mortgage Corporation).

For all loans sold through the "cash window," the interest rate at which the loan will be priced and therefore discounted is contractually locked at the time of sale, rather than at the time of delivery of title to the loan. This contractual discount rate is called the required net yield. The lender that is selling forward for four or eight weeks has transferred the net-yield-interest-rate risk (that it will fluctuate in the secondary market) to Farmer Mac. For any cash sale, since the loan already exists, its interest rate is known and the sale is likely to generate the par, or full price, from Farmer Mac. The full value of the loan will be received as long as the loan's interest rate is at least as high as the required net yield. But for the forward sale, the interest rate of the loan is only set if the lender has received the borrower's commitment to the loan at a specific interest rate. So the lender may prefer that the borrower commit to a specific interest rate, that is, to lock in the rate in the primary or loan-origination market, then sell the loan forward to Farmer Mac, leaving the remaining four or eight weeks for completion of the loan process. The advantage of this approach is like the cash sale of an existing portfolio loan, lenders can be fairly certain of receiving the par price by setting the loan interest rate at least as high as Farmer Mac's net yield.

Farmer Mac provides daily interest rates, that is, net yields, at its web site [www.farmermac.com]. A glance at any day's quotations for either the Farmer Mac I or II programs reveals that for longer forward loan sales, the required net yield tends to be higher. This is due to the interest-rate risk Farmer Mac is accepting by locking in the net yield. It still must sell its AMBSs through Wall Street when it finally has the loans as collateral. Interest rates may change over the four or eight weeks prior to delivery of title to the loans, so Farmer Mac risks an increase in credit market and MBS interest rates, making the loan pool less valuable, and causing it to suffer a monetary loss when it finally sells the AMBSs. Like other institutions involved in credit-market transactions, Farmer Mac has devised ways to hedge away this interest-rate risk.[8]

The lender working with Farmer Mac I or II program borrowers has three alternative means of setting the loan interest rate for the borrower. The first method is to sell the loan forward to Farmer Mac four or eight weeks while simultaneously locking the loan interest rate with the borrower (assuming the borrower agrees to the rate). This eliminates all interest-rate risk for the lender, but is more costly to the borrower due to higher net yields for longer forward loan sales, and there remains a chance of borrower fallout, especially if interest rates drop after the loan rate is locked. A second approach is to let the loan process near completion, and set the loan interest rate at the end of the process,

when the net yield to Farmer Mac is determined from its seven-day delivery figures. Once again this eliminates interest-rate risk, but borrowers can be frus-
trated by not knowing the loan's actual interest rate until the end, and this is not
as "user-friendly" a means of locking the loan interest rate. Borrowers may prefer to deal with another lender that is willing to provide the actual rate early in
the loan process.

The third method of setting the rate is to lock the borrower's loan interest
rate at the beginning of the loan process, four or more weeks before the loan
can be closed, while letting the sale of the loan float until it is ready to close
and be delivered to Farmer Mac. This has the advantage of keeping the borrower happy by guaranteeing him or her a rate, but creates the problem of
interest-rate risk for the lender's sale of the loan to Farmer Mac, due to the possibility of rising rates while the loan is in process. If interest rates are on the
rise, the borrower will be strongly motivated to close the loan due to his or her
earlier, lower guaranteed rate. But the lender will not get full or par value for
the loan due to its below-net-yield interest rate when it is finally sold. It will be
discounted by Farmer Mac. What can the lender do to deal with this secondary
market interest rate risk?

Since the risk is that the value of the closed loan is falling due to rising
interest rates, the lender needs to hedge, or take a position in another security
whose price responds in the opposite way to that of the mortgages, by trading
some other interest-rate-sensitive instrument. The trade the lender needs to
make should be in a security whose value will rise as the loan value is declining during the run-up in interest rates. Ordinary Treasury Bills and Bonds will
suffer declining values when credit-market interest rates rise. But interest-rate
futures contracts allow the trader to take a position, that is, to exercise a trade,
that will become more valuable as interest rates rise. Similar to going long on a
commodities contract, as the commodity becomes more valuable due to market
factors over time, the contract becomes worth more as time goes by. In the case
of interest-rate futures contracts, as interest rates rise, the contract is worth less
and less. So rather than go long, the lender should sell short. By selling short an
interest-rate futures contract, the lender will be receiving his or her price for the
contract at the beginning of the trade, when its value is highest (because interest rates are lowest—remember, interest rates and securities' values are
inversely related). Then, when the value of the loan to which it has committed
is declining, the value of its futures contract trade will be increasing, both due
to rising interest rates, thus offsetting the decline in loan value—a hedge.

A number of interest-rate futures contracts have been utilized by loan
originators to hedge for primary market interest-rate risk. One popular contract
is the Ten-Year Treasury Note Futures Contract traded at the Chicago Board of
Trade. The contract allows the long-position trader to receive a $100,000, 8%
semiannual coupon Treasury Note, or its equivalent, at maturity of the contract

for whatever price the long trader paid when he or she purchased the contract. Our lender, who sold short, would have received that initial payment. (Actually, a minimum amount would be required by the short trader as an investment in his or her account, then the account would be credited each day with any change in value of the contract.) These T-Note Futures Contracts are so popular that in 1999 there was approximately fifty-six billion dollars in value of contracts outstanding! So selling short interest-rate futures contracts is one way to hedge for rising secondary market rates under the third method of setting the loan rate for the borrower.

A second method of hedging for rising interest rates is to trade in interest-rate futures options. An **option contract** is a right but not an obligation of the contract holder (the long) to buy/call or sell/put another security at a guaranteed exercise or strike price, during the option period. At the turn of the millennium, there are option contracts for stock, stock indexes, bond securities, other indexes, and futures contracts. Interest-rate futures options give the long position the right but not the obligation to buy/call or sell/put an interest-rate futures contract, such as the Ten-Year Treasury Note Contract at the strike price during the option period. For a lender facing the risk of rising secondary market interest rates and, hence, a loss on the sale of his or her lower-interest-rate loan (because its rate was locked for the borrower earlier, at the start of the four- or eight-week loan process), the correct interest-rate futures option trade is that of a long-put. A long-put means the lender would then have the right but not the obligation to sell an interest-rate futures contract to the other party to the transaction, the short, at a guaranteed strike price. This right to sell becomes more valuable as the market value of the futures contract falls in a rising rate environment. In effect, the long trader could buy a futures contract cheaply in the futures market, then turn around and sell it to the short trader at the strike price for a guaranteed profit. So, the two rules of thumb for lenders attempting to hedge their loan originations against rising secondary market interest rates are (1) sell short interest-rate futures contracts and (2) trade long-put interest-rate futures options contracts.

For both Farmer Mac I and II loan purchases, acceptable loans include adjustable-rate and fixed-rate mortgages. Popular loan terms are seven and fifteen years for the fixed-rate loans. Adjustables come in five-, ten-, fifteen-, and twenty-five-year terms, with U.S. Treasury Security interest rates and business prime rates being acceptable as indexes. Reset mortgages, those that fix the interest rate for five years and then are either paid off as a balloon payment or adjusted to current market rates, are available under Farmer Mac II guidelines. Most loans do include prepayment penalties to protect the AMBS investor, and most loans require the servicing of the originator, who collects a small fee with each loan payment received. In the near future, Farmer Mac may install an on-line secondary-market purchasing program; for the time being, loan sales are transacted by telephone between approved lenders and the Washington, D.C. headquarters.

Farmer Mac's Ag Vantage Program differs significantly from its two AMBS programs. While Farmer Mac I and II are both asset sales of loans by ag lenders to Farmer Mac, or a swap of loans for its AMBSs, Ag Vantage is instead a bond sale from the lender to Farmer Mac, with the lender using approved loans as collateral for the bond. The most obvious difference between the two forms of mortgage-backed securities is that I and II are asset sales, while Ag Vantage is a debt offering. Therefore, while lenders sell or swap loans under Farmer Mac I and II and their loans go off balance sheet, with Ag Vantage the loans remain in portfolio. In effect, both sides of the balance sheet grow under the Ag Vantage Program. The lender receives an inflow of cash from Farmer Mac at the time of sale, then lends these funds out to repeat the loan process with new customers. One obvious drawback of this bond sale as opposed to an asset sale is in the lender's capital limitations. Highly regulated bank lenders are limited as to the balance sheet debt they can incur and still meet capital guidelines under recent federal legislation, including the 1989 Financial Institutions Reform, Recovery and Enforcement Act and the 1991 Federal Deposit Insurance Improvement Act.

Terms under the Ag Vantage Program are from thirty days to ten years, with interest rates about 0.5% above U.S. Treasury interest rates of similar maturity. Inherent in this program is the effect such borrowing will have on the lender's loan interest rates. Such a high cost of borrowing in the secondary market will force lender–bond sellers to set the rates to their loan borrowers at some reasonable spread over this cost of funds. Like the capital limitations cited above, this appears to limit the usefulness of the Ag Vantage program when compared to Farmer Mac's other AMBS programs.

Although four years ago there was some question as to whether Farmer Mac would survive as the agricultural industry's secondary-market connection for real-estate loans, since the Farm Credit System Reform Act of 1996, it has stood on much firmer financial ground. Although its stockholder's equity had fallen below the statutory minimum of $25 million in 1995, it now stands at near $80 million. Personnel grew from a staff of fourteen to twenty-eight in 1999. Outstanding AMBSs now total $1.4 billion, while in 1996 they totaled just $827 million. Most important, this GSE's (Government Sponsored Enterprise) operations are now streamlined, with more technology likely to follow in its secondary-market operations. Its stock is now listed on the big board, the New York Stock Exchange, under the ticker symbol of *AGM*. The dollar value of purchased or guaranteed loans was up sharply in 1999 over 1998. Farmer Mac's net interest income increased in the same time period as well. Wall Street has embraced its stock with the price more than doubling this past year. It looks as if Farmer Mac is here to stay, and has begun establishing the ag mortgage secondary market it was chartered by Congress to provide.

■ Private Financing

Similar to the residential mortgage industry, private financing for agricultural property tends to be family loans, seller financing, and broker-arranged mortgages where the borrower fails to meet financial institution underwriting criteria and must seek "hard money." Seller financing most often is suitable when the farm property has been paid off by the seller during his or her tenure, so it is owned free and clear at the time of sale. These transactions are set up during the property escrow, the responsibility for the loan documents falling to the escrow or title officer handling the deeding and other escrow paperwork. One advantage of seller financing is that no minimum down payment is required—it is at the discretion of the seller. Sellers generally know the value of the property better than other parties and can assess whether the loan-to-value ratio that their financing will create is excessive. This form of mortgage lending can also be used within family transactions where the seller does not want to deed or bequeath the property outright to the younger family member such as a nephew or niece deciding to buy the family farm.

Another common use of seller financing is in suburban development, where a city is surrounded by agricultural property, which has now become more valuable as a residential or commercial subdivision. The original landowners may find that continuing to farm just doesn't pay; that is, there is too high an opportunity cost to operating the family farm when it could be sold to developers for a handsome capital gain. Developers face particularly difficult financing hurdles due to the high-risk nature of development, their relatively small operating capital, the lack of any initial collateral other than the land, and the often difficult political process of rezoning and subdividing property. Farm-seller financing for development often includes a promissory note and mortgage contractual clause that most developers request of the seller, the **subordination clause**. This statement in a promissory note requires the seller/lender to allow other institutional lenders to hold a higher-security claim on the property if they provide subsequent financing should there be a default by the developer/borrower. Without the subordination clause, institutional lenders would be loathe to provide construction and development financing.

Amortization patterns for private ag financing vary considerably. Interest-only loans requiring a balloon payment are popular, as are fully amortizing long-term fixed-rate loans. In escrow, the loan terms can be tailored to meet the needs of both parties. For example, the seller may allow the borrower to make smaller payments in the first several years of the mortgage, while the property is raised to a more productive level, after which the loan can be fully amortizing.

Brokered loans involve a mortgage broker or real-estate broker who negotiates the loan from a private investor for an ag borrower, receiving a commission of several points for the effort. One point is 1% of the loan amount and loan points for private financing can range from two or three up to ten or more in tight-money times. Presently, private money brokered loans are a very small part of total loan transactions, and points are toward the low end of the range.

From the buyer/borrower standpoint, private financing can be a boon to the farm purchase. In addition to the formal mortgage or deed-of-trust financing, sellers can create the **land contract** or **contract for deed**, which is a form of installment financing similar to an automobile loan. Title passes when the promissory note is paid in full. This and other private financing can be the viable means of acquiring the farm property if the borrower or the property does not meet institutional guidelines. In the past, one problem surfaced with the land contract, that of equitable title. Civil courts tend to view the installment payments as partial purchase of the property, so that if the buyer/borrower defaults on the promissory after some time has elapsed, he or she is entitled to partial equity in the property, making foreclosure more difficult.

With the ongoing effort of the Farm Service Agency to target young and first-time buyers, private loans should remain a minor factor in ag financing, until such time as the Federal Reserve Bank institutes a tight money policy and institutional money is once again hard to obtain.

■ Beyond the Year Two Thousand

As we enter the twenty-first century, agribusiness and its financing are changing. Farms and other ag producers become larger and more automated, and are developing contractual ties between themselves and processors. The providers of ag financing are becoming more sophisticated as well, with improved communication technology through the use of computers aiding underwriting and loan-specific software speeding processing and marketing of the loan portfolio. In the secondary market, Wall Street and other credit markets are playing an increasing role, while public agency guidelines are revised through legislation. No longer are there twelve district Farm Credit Banks, and Farmers' Home Administration has been replaced by the Farm Service Agency. What other trends or changes do the authors think plausible for the first decade of the third millennium?

From 1989 to 1996, the share of ag debt provided by the Farm Credit System held steady near 26% of the total, the USDA/FSA percentage of the total dropped by 7%, life insurance holdings held steady at 6%, and commercial bank ag debt increased from 33% to 39% of the total.[9] Competition has intensified as the secondary market has become viable, loan sizes are on the increase, and nontraditional sources such as foreign investors are on the rise. Though the Farm Credit System's outstanding loan volume is trending upward, declines in loans outstanding from 1991 to 1992 and 1995 to 1996 hint at its increased competition. These trends in ag lender debt holdings should continue in the near future.

Government Sponsored Enterprises (taxpayer linked) such as the Farm Services Agency and Farmer Mac may come under political scrutiny from time to time, but have stood the test of competition during the twentieth century. The GSE's viability depends on the competition. Although commercial banks are the most highly regulated of the agribusiness lenders, theirs is a sophisticated industry, where savvy management has effected many successful mergers in the 1990s, and profitability is the key concern. Banks' share of ag debt will likely continue to increase, with Farmer Mac and the secondary market giving key banking assistance.

As banks continue to provide additional services to customers and focus on "relationship banking," the small farm or other agribusiness owner is likely to find the banker helping with matters beyond the checking account and a real-estate or operating loan. With the erosion of Glass-Steagall, banks now provide individual retirement planning (IRA accounts), estate management, and other financial planning services such as employee retirement and cash budgeting. With the progress in technology, these services are relatively easy to provide to agribusiness borrowers. Leasing should continue to erode the purchase loan market for intermediate-term assets such as farm equipment. Though leases don't provide significant advantages over the long haul (see chapter 4), their start-up feature of little down payment and tax deductibility of rent will keep them viable.

For the new millennium, one thing is clear. The agribusiness borrower will continue to have his or her financial needs met by a variety of providers, including GSEs, private money sources, commercial banks, and other financial institutions.

■ Questions and Problems

9-1. Smith's Irrigation Supply, Inc., purchases PVC pipe from the manufacturer under credit terms of 3/15, n30. Assuming it buys $500/day, 360 days per year, answer the following:

    a. If Smith's pays on the tenth day, how much is its total discount for the year?

    b. Assume instead that Smith's forgoes the early payment discount for the extra financing it receives. What is the effective interest rate (cost) of the forgone early discount?

9-2. Farmer Mac allows a seven-day, four-week, or eight-week delivery of loans it purchases under its Farmer Mac I and II programs. Explain its pricing relative to its interest-rate risk for these different delivery periods. (Hint: how does its *required net yield* differ for the three periods?)

9-3. What is loan "fallout"? What typically causes fallout to occur?

9-4. The third method of locking the interest rates for a lender who sells loans to Farmer Mac is to provide a locked rate to the borrower soon after application but allow the sale interest rate (Farmer Mac's net yield) to float until the loan process nears completion. This method creates interest-rate risk for the lender. Explain how the lender can hedge this interest-rate risk using interest-rate futures contracts and interest-rate futures options contracts. Which trade should it initiate at the start of the loan process with each type of contract?

9-5. Bill and Susan Kellogg have decided to expand their poultry operation and have lined up private financing through an ag real-estate broker. Terms of the financing:

Loan amount	$100,000
Term	7 years
Interest rate	9%
Monthly payments	
Points charged by broker	2 points

a. How much are the points in dollars?

b. How much is each monthly payment?

c. At the end of two years, the loan is sold by the original private lender to another investor. The investor requires a yield of 10%. How much does the seller receive (assume there are no fees involved)?

■ Notes

1. Jack Nowakowski, American Council of Life Insurance, August 1999.
2. American Council of Life Insurance.
3. U.S. Department of Agriculture, *Situation and Outlook Report,* February 1999.
4. USDA, *Situation and Outlook Report,* 1999.
5. USDA, *Situation and Outlook Report,* 1999.
6. USDA, *Situation and Outlook Report,* 1999.
7. USDA, *Situation and Outlook Report,* 1999.
8. Mary Maloney, Farmer Mac, July 1999.
9. U.S. Department of Agriculture, *Situation and Outlook Report,* February 1997.

Appendix

Table A1. Compound interest (future value of a dollar)

years	3%	4%	5%	6%	7%	8%	9%	10%	11%	12%	13%	14%	15%
1	1.03	1.04	1.05	1.06	1.07	1.08	1.09	1.1	1.11	1.12	1.13	1.14	1.15
2	1.0609	1.0816	1.1025	1.1236	1.1449	1.1664	1.1881	1.21	1.2321	1.2544	1.2769	1.2996	1.3225
3	1.0927	1.1249	1.1576	1.191	1.225	1.2597	1.295	1.331	1.3676	1.4049	1.4429	1.4815	1.5209
4	1.1255	1.1699	1.2155	1.2625	1.3108	1.3605	1.4116	1.4641	1.5181	1.5735	1.6305	1.689	1.749
5	1.1593	1.2167	1.2763	1.3382	1.4026	1.4693	1.5386	1.6105	1.6851	1.7623	1.8424	1.9254	2.0114
6	1.1941	1.2653	1.3401	1.4185	1.5007	1.5869	1.6771	1.7716	1.8704	1.9738	2.082	2.195	2.3131
7	1.2299	1.3159	1.4071	1.5036	1.6058	1.7138	1.828	1.9487	2.0762	2.2107	2.3526	2.5023	2.66
8	1.2668	1.3686	1.4775	1.5938	1.7182	1.8509	1.9926	2.1436	2.3045	2.476	2.6584	2.8526	3.059
9	1.3048	1.4233	1.5513	1.6895	1.8385	1.999	2.1719	2.3579	2.558	2.7731	3.004	3.2519	3.5179
10	1.3439	1.4802	1.6289	1.7908	1.9672	2.1589	2.3674	2.5937	2.8394	3.1058	3.3946	3.7072	4.0456
11	1.3842	1.5395	1.7103	1.8983	2.1049	2.3316	2.5804	2.8531	3.1518	3.4785	3.8359	4.2262	4.6524
12	1.4258	1.601	1.7959	2.0122	2.2522	2.5182	2.8127	3.1384	3.4985	3.896	4.3345	4.8179	5.3503
13	1.4685	1.6651	1.8856	2.1329	2.4098	2.7196	3.0658	3.4523	3.8833	4.3635	4.898	5.4924	6.1528
14	1.5126	1.7317	1.9799	2.2609	2.5785	2.9372	3.3417	3.7975	4.3104	4.8871	5.5348	6.2613	7.0757
15	1.558	1.8009	2.0789	2.3966	2.759	3.1722	3.6425	4.1772	4.7846	5.4736	6.2543	7.1379	8.1371
16	1.6047	1.873	2.1829	2.5404	2.9522	3.4259	3.9703	4.595	5.3109	6.1304	7.0673	8.1372	9.3576
17	1.6528	1.9479	2.292	2.6928	3.1588	3.7	4.3276	5.0545	5.8951	6.866	7.9861	9.2765	10.761
18	1.7024	2.0258	2.4066	2.8543	3.3799	3.996	4.7171	5.5599	6.5436	7.69	9.0243	10.575	12.375
19	1.7535	2.1068	2.527	3.0256	3.6165	4.3157	5.1417	6.1159	7.2633	8.6128	10.197	12.056	14.232
20	1.8061	2.1911	2.6533	3.2071	3.8697	4.661	5.6044	6.7275	8.0623	9.6463	11.523	13.743	16.367
21	1.8603	2.2788	2.786	3.3996	4.1406	5.0338	6.1088	7.4002	8.9492	10.804	13.021	15.668	18.882

22	1.9161	2.3699	2.9253	3.6035	4.4304	5.4365	6.6586	8.1403	9.9336	12.1	14.714	17.861	21.645
23	1.9736	2.4647	3.0715	3.8197	4.7405	5.8715	7.2579	8.9543	11.026	13.552	16.627	20.362	24.891
24	2.0328	2.5633	3.2251	4.0489	5.0724	6.3412	7.9111	9.8497	12.239	15.179	18.788	23.212	28.625
25	2.0938	2.6658	3.3864	4.2919	5.4274	6.8485	8.6231	10.835	13.585	17	21.231	26.462	32.919
26	2.1566	2.7725	3.5557	4.5494	5.8074	7.3964	9.3992	11.918	15.08	19.04	23.991	30.167	37.857
27	2.2213	2.8834	3.7335	4.8223	6.2139	7.9881	10.245	13.11	16.739	21.325	27.109	34.39	43.535
28	2.2879	2.9987	3.9201	5.1117	6.6488	8.6271	11.167	14.421	18.58	23.884	30.633	39.204	50.066
29	2.3566	3.1187	4.1161	5.4184	7.1143	9.3173	12.172	15.863	20.624	26.75	34.616	44.693	57.575
30	2.4273	3.2434	4.3219	5.7435	7.6123	10.063	13.268	17.449	22.892	29.96	39.116	50.95	66.212
31	2.5001	3.3731	4.538	6.0881	8.1451	10.868	14.462	19.194	25.41	33.555	44.201	58.083	76.144
32	2.5751	3.5081	4.7649	6.4534	8.7153	11.737	15.763	21.114	28.206	37.582	49.947	66.215	87.565
33	2.6523	3.6484	5.0032	6.8406	9.3253	12.676	17.182	23.225	31.308	42.092	56.44	75.485	100.7
34	2.7319	3.7943	5.2533	7.251	9.9781	13.69	18.728	25.548	34.752	47.143	63.777	86.053	115.8
35	2.8139	3.9461	5.516	7.6861	10.677	14.785	20.414	28.102	38.575	52.8	72.069	98.1	133.18
36	2.8983	4.1039	5.7918	8.1473	11.424	15.968	22.251	30.913	42.818	59.136	81.437	111.83	153.15
37	2.9852	4.2681	6.0814	8.6361	12.224	17.246	24.254	34.004	47.528	66.232	92.024	127.49	176.12
38	3.0748	4.4388	6.3855	9.1543	13.079	18.625	26.437	37.404	52.756	74.18	103.99	145.34	202.54
39	3.167	4.6164	6.7048	9.7035	13.995	20.115	28.816	41.145	58.559	83.081	117.51	165.69	232.92
40	3.262	4.801	7.04	10.286	14.974	21.725	31.409	45.259	65.001	93.051	132.78	188.88	267.86
41	3.3599	4.9931	7.392	10.903	16.023	23.462	34.236	49.785	72.151	104.22	150.04	215.33	308.04
42	3.4607	5.1928	7.7616	11.557	17.144	25.339	37.318	54.764	80.088	116.72	169.55	245.47	354.25
43	3.5645	5.4005	8.1497	12.25	18.344	27.367	40.676	60.24	88.897	130.73	191.59	279.84	407.39
44	3.6715	5.6165	8.5572	12.985	19.628	29.556	44.337	66.264	98.676	146.42	216.5	319.02	468.5
45	3.7816	5.8412	8.985	13.765	21.002	31.92	48.327	72.89	109.53	163.99	244.64	363.68	538.77

Table A2. Present value of a dollar

years	3%	4%	5%	6%	7%	8%	9%	10%	11%	12%	13%	14%	15%
1	0.9709	0.9615	0.9524	0.9434	0.9346	0.9259	0.9174	0.9091	0.9009	0.8929	0.885	0.8772	0.8696
2	0.9426	0.9246	0.907	0.89	0.8734	0.8573	0.8417	0.8264	0.8116	0.7972	0.7831	0.7695	0.7561
3	0.9151	0.889	0.8638	0.8396	0.8163	0.7938	0.7722	0.7513	0.7312	0.7118	0.6931	0.675	0.6575
4	0.8885	0.8548	0.8227	0.7921	0.7629	0.735	0.7084	0.683	0.6587	0.6355	0.6133	0.5921	0.5718
5	0.8626	0.8219	0.7835	0.7473	0.713	0.6806	0.6499	0.6209	0.5935	0.5674	0.5428	0.5194	0.4972
6	0.8375	0.7903	0.7462	0.705	0.6663	0.6302	0.5963	0.5645	0.5346	0.5066	0.4803	0.4556	0.4323
7	0.8131	0.7599	0.7107	0.6651	0.6227	0.5835	0.547	0.5132	0.4817	0.4523	0.4251	0.3996	0.3759
8	0.7894	0.7307	0.6768	0.6274	0.582	0.5403	0.5019	0.4665	0.4339	0.4039	0.3762	0.3506	0.3269
9	0.7664	0.7026	0.6446	0.5919	0.5439	0.5002	0.4604	0.4241	0.3909	0.3606	0.3329	0.3075	0.2843
10	0.7441	0.6756	0.6139	0.5584	0.5083	0.4632	0.4224	0.3855	0.3522	0.322	0.2946	0.2697	0.2472
11	0.7224	0.6496	0.5847	0.5268	0.4751	0.4289	0.3875	0.3505	0.3173	0.2875	0.2607	0.2366	0.2149
12	0.7014	0.6246	0.5568	0.497	0.444	0.3971	0.3555	0.3186	0.2858	0.2567	0.2307	0.2076	0.1869
13	0.681	0.6006	0.5303	0.4688	0.415	0.3677	0.3262	0.2897	0.2575	0.2292	0.2042	0.1821	0.1625
14	0.6611	0.5775	0.5051	0.4423	0.3878	0.3405	0.2992	0.2633	0.232	0.2046	0.1807	0.1597	0.1413
15	0.6419	0.5553	0.481	0.4173	0.3624	0.3152	0.2745	0.2394	0.209	0.1827	0.1599	0.1401	0.1229
16	0.6232	0.5339	0.4581	0.3936	0.3387	0.2919	0.2519	0.2176	0.1883	0.1631	0.1415	0.1229	0.1069
17	0.605	0.5134	0.4363	0.3714	0.3166	0.2703	0.2311	0.1978	0.1696	0.1456	0.1252	0.1078	0.0929
18	0.5874	0.4936	0.4155	0.3503	0.2959	0.2502	0.212	0.1799	0.1528	0.13	0.1108	0.0946	0.0808
19	0.5703	0.4746	0.3957	0.3305	0.2765	0.2317	0.1945	0.1635	0.1377	0.1161	0.0981	0.0829	0.0703
20	0.5537	0.4564	0.3769	0.3118	0.2584	0.2145	0.1784	0.1486	0.124	0.1037	0.0868	0.0728	0.0611
21	0.5375	0.4388	0.3589	0.2942	0.2415	0.1987	0.1637	0.1351	0.1117	0.0926	0.0768	0.0638	0.0531

22	0.5219	0.422	0.3418	0.2775	0.2257	0.1839	0.1502	0.1228	0.1007	0.0826	0.068	0.056	0.0462
23	0.5067	0.4057	0.3256	0.2618	0.2109	0.1703	0.1378	0.1117	0.0907	0.0738	0.0601	0.0491	0.0402
24	0.4919	0.3901	0.3101	0.247	0.1971	0.1577	0.1264	0.1015	0.0817	0.0659	0.0532	0.0431	0.0349
25	0.4776	0.3751	0.2953	0.233	0.1842	0.146	0.116	0.0923	0.0736	0.0588	0.0471	0.0378	0.0304
26	0.4637	0.3607	0.2812	0.2198	0.1722	0.1352	0.1064	0.0839	0.0663	0.0525	0.0417	0.0331	0.0264
27	0.4502	0.3468	0.2678	0.2074	0.1609	0.1252	0.0976	0.0763	0.0597	0.0469	0.0369	0.0291	0.023
28	0.4371	0.3335	0.2551	0.1956	0.1504	0.1159	0.0895	0.0693	0.0538	0.0419	0.0326	0.0255	0.02
29	0.4243	0.3207	0.2429	0.1846	0.1406	0.1073	0.0822	0.063	0.0485	0.0374	0.0289	0.0224	0.0174
30	0.412	0.3083	0.2314	0.1741	0.1314	0.0994	0.0754	0.0573	0.0437	0.0334	0.0256	0.0196	0.0151
31	0.4	0.2965	0.2204	0.1643	0.1228	0.092	0.0691	0.0521	0.0394	0.0298	0.0226	0.0172	0.0131
32	0.3883	0.2851	0.2099	0.155	0.1147	0.0852	0.0634	0.0474	0.0355	0.0266	0.02	0.0151	0.0114
33	0.377	0.2741	0.1999	0.1462	0.1072	0.0789	0.0582	0.0431	0.0319	0.0238	0.0177	0.0132	0.0099
34	0.366	0.2636	0.1904	0.1379	0.1002	0.073	0.0534	0.0391	0.0288	0.0212	0.0157	0.0116	0.0086
35	0.3554	0.2534	0.1813	0.1301	0.0937	0.0676	0.049	0.0356	0.0259	0.0189	0.0139	0.0102	0.0075
36	0.345	0.2437	0.1727	0.1227	0.0875	0.0626	0.0449	0.0323	0.0234	0.0169	0.0123	0.0089	0.0065
37	0.335	0.2343	0.1644	0.1158	0.0818	0.058	0.0412	0.0294	0.021	0.0151	0.0109	0.0078	0.0057
38	0.3252	0.2253	0.1566	0.1092	0.0765	0.0537	0.0378	0.0267	0.019	0.0135	0.0096	0.0069	0.0049
39	0.3158	0.2166	0.1491	0.1031	0.0715	0.0497	0.0347	0.0243	0.0171	0.012	0.0085	0.006	0.0043
40	0.3066	0.2083	0.142	0.0972	0.0668	0.046	0.0318	0.0221	0.0154	0.0107	0.0075	0.0053	0.0037
41	0.2976	0.2003	0.1353	0.0917	0.0624	0.0426	0.0292	0.0201	0.0139	0.0096	0.0067	0.0046	0.0032
42	0.289	0.1926	0.1288	0.0865	0.0583	0.0395	0.0268	0.0183	0.0125	0.0086	0.0059	0.0041	0.0028
43	0.2805	0.1852	0.1227	0.0816	0.0545	0.0365	0.0246	0.0166	0.0112	0.0076	0.0052	0.0036	0.0025
44	0.2724	0.178	0.1169	0.077	0.0509	0.0338	0.0226	0.0151	0.0101	0.0068	0.0046	0.0031	0.0021
45	0.2644	0.1712	0.1113	0.0727	0.0476	0.0313	0.0207	0.0137	0.0091	0.0061	0.0041	0.0027	0.0019

Table A3. Amortization

years	3%	4%	5%	6%	7%	8%	9%	10%	11%	12%	13%	14%	15%
1	1.03	1.04	1.05	1.06	1.07	1.08	1.09	1.1	1.11	1.12	1.13	1.14	1.15
2	0.5226	0.5302	0.5378	0.5454	0.5531	0.5608	0.5685	0.5762	0.5839	0.5917	0.5995	0.6073	0.6151
3	0.3535	0.3603	0.3672	0.3741	0.3811	0.388	0.3951	0.4021	0.4092	0.4163	0.4235	0.4307	0.438
4	0.269	0.2755	0.282	0.2886	0.2952	0.3019	0.3087	0.3155	0.3223	0.3292	0.3362	0.3432	0.3503
5	0.2184	0.2246	0.231	0.2374	0.2439	0.2505	0.2571	0.2638	0.2706	0.2774	0.2843	0.2913	0.2983
6	0.1846	0.1908	0.197	0.2034	0.2098	0.2163	0.2229	0.2296	0.2364	0.2432	0.2502	0.2572	0.2642
7	0.1605	0.1666	0.1728	0.1791	0.1856	0.1921	0.1987	0.2054	0.2122	0.2191	0.2261	0.2332	0.2404
8	0.1425	0.1485	0.1547	0.161	0.1675	0.174	0.1807	0.1874	0.1943	0.2013	0.2084	0.2156	0.2229
9	0.1284	0.1345	0.1407	0.147	0.1535	0.1601	0.1668	0.1736	0.1806	0.1877	0.1949	0.2022	0.2096
10	0.1172	0.1233	0.1295	0.1359	0.1424	0.149	0.1558	0.1627	0.1698	0.177	0.1843	0.1917	0.1993
11	0.1081	0.1141	0.1204	0.1268	0.1334	0.1401	0.1469	0.154	0.1611	0.1684	0.1758	0.1834	0.1911
12	0.1005	0.1066	0.1128	0.1193	0.1259	0.1327	0.1397	0.1468	0.154	0.1614	0.169	0.1767	0.1845
13	0.094	0.1001	0.1065	0.113	0.1197	0.1265	0.1336	0.1408	0.1482	0.1557	0.1634	0.1712	0.1791
14	0.0885	0.0947	0.101	0.1076	0.1143	0.1213	0.1284	0.1357	0.1432	0.1509	0.1587	0.1666	0.1747
15	0.0838	0.0899	0.0963	0.103	0.1098	0.1168	0.1241	0.1315	0.1391	0.1468	0.1547	0.1628	0.171
16	0.0796	0.0858	0.0923	0.099	0.1059	0.113	0.1203	0.1278	0.1355	0.1434	0.1514	0.1596	0.1679
17	0.076	0.0822	0.0887	0.0954	0.1024	0.1096	0.117	0.1247	0.1325	0.1405	0.1486	0.1569	0.1654
18	0.0727	0.079	0.0855	0.0924	0.0994	0.1067	0.1142	0.1219	0.1298	0.1379	0.1462	0.1546	0.1632
19	0.0698	0.0761	0.0827	0.0896	0.0968	0.1041	0.1117	0.1195	0.1276	0.1358	0.1441	0.1527	0.1613
20	0.0672	0.0736	0.0802	0.0872	0.0944	0.1019	0.1095	0.1175	0.1256	0.1339	0.1424	0.151	0.1598
21	0.0649	0.0713	0.078	0.085	0.0923	0.0998	0.1076	0.1156	0.1238	0.1322	0.1408	0.1495	0.1584

22	0.0627	0.0692	0.076	0.083	0.0904	0.098	0.1059	0.114	0.1223	0.1308	0.1395	0.1483	0.1573
23	0.0608	0.0673	0.0741	0.0813	0.0887	0.0964	0.1044	0.1126	0.121	0.1296	0.1383	0.1472	0.1563
24	0.059	0.0656	0.0725	0.0797	0.0872	0.095	0.103	0.1113	0.1198	0.1285	0.1373	0.1463	0.1554
25	0.0574	0.064	0.071	0.0782	0.0858	0.0937	0.1018	0.1102	0.1187	0.1275	0.1364	0.1455	0.1547
26	0.0559	0.0626	0.0696	0.0769	0.0846	0.0925	0.1007	0.1092	0.1178	0.1267	0.1357	0.1448	0.1541
27	0.0546	0.0612	0.0683	0.0757	0.0834	0.0914	0.0997	0.1083	0.117	0.1259	0.135	0.1442	0.1535
28	0.0533	0.06	0.0671	0.0746	0.0824	0.0905	0.0989	0.1075	0.1163	0.1252	0.1344	0.1437	0.1531
29	0.0521	0.0589	0.066	0.0736	0.0814	0.0896	0.0981	0.1067	0.1156	0.1247	0.1339	0.1432	0.1527
30	0.051	0.0578	0.0651	0.0726	0.0806	0.0888	0.0973	0.1061	0.115	0.1241	0.1334	0.1428	0.1523
31	0.05	0.0569	0.0641	0.0718	0.0798	0.0881	0.0967	0.1055	0.1145	0.1237	0.133	0.1425	0.152
32	0.049	0.0559	0.0633	0.071	0.0791	0.0875	0.0961	0.105	0.114	0.1233	0.1327	0.1421	0.1517
33	0.0482	0.0551	0.0625	0.0703	0.0784	0.0869	0.0956	0.1045	0.1136	0.1229	0.1323	0.1419	0.1515
34	0.0473	0.0543	0.0618	0.0696	0.0778	0.0863	0.0951	0.1041	0.1133	0.1226	0.1321	0.1416	0.1513
35	0.0465	0.0536	0.0611	0.069	0.0772	0.0858	0.0946	0.1037	0.1129	0.1223	0.1318	0.1414	0.1511
36	0.0458	0.0529	0.0604	0.0684	0.0767	0.0853	0.0942	0.1033	0.1126	0.1221	0.1316	0.1413	0.151
37	0.0451	0.0522	0.0598	0.0679	0.0762	0.0849	0.0939	0.103	0.1124	0.1218	0.1314	0.1411	0.1509
38	0.0445	0.0516	0.0593	0.0674	0.0758	0.0845	0.0935	0.1027	0.1121	0.1216	0.1313	0.141	0.1507
39	0.0438	0.0511	0.0588	0.0669	0.0754	0.0842	0.0932	0.1025	0.1119	0.1215	0.1311	0.1409	0.1506
40	0.0433	0.0505	0.0583	0.0665	0.075	0.0839	0.093	0.1023	0.1117	0.1213	0.131	0.1407	0.1506
41	0.0427	0.05	0.0578	0.0661	0.0747	0.0836	0.0927	0.102	0.1115	0.1212	0.1309	0.1407	0.1505
42	0.0422	0.0495	0.0574	0.0657	0.0743	0.0833	0.0925	0.1019	0.1114	0.121	0.1308	0.1406	0.1504
43	0.0417	0.0491	0.057	0.0653	0.074	0.083	0.0923	0.1017	0.1113	0.1209	0.1307	0.1405	0.1504
44	0.0412	0.0487	0.0566	0.065	0.0738	0.0828	0.0921	0.1015	0.1111	0.1208	0.1306	0.1404	0.1503
45	0.0408	0.0483	0.0563	0.0647	0.0735	0.0826	0.0919	0.1014	0.111	0.1207	0.1305	0.1404	0.1503

Table A4. Future value of an ordinary annuity

years	3%	4%	5%	6%	7%	8%	9%	10%	11%	12%	13%	14%	15%
1	1	1	1	1	1	1	1	1	1	1	1	1	1
2	2.03	2.04	2.05	2.06	2.07	2.08	2.09	2.1	2.11	2.12	2.13	2.14	2.15
3	3.0909	3.1216	3.1525	3.1836	3.2149	3.2464	3.2781	3.31	3.3421	3.3744	3.4069	3.4396	3.4725
4	4.1836	4.2465	4.3101	4.3746	4.4399	4.5061	4.5731	4.641	4.7097	4.7793	4.8498	4.9211	4.9934
5	5.3091	5.4163	5.5256	5.6371	5.7507	5.8666	5.9847	6.1051	6.2278	6.3528	6.4803	6.6101	6.7424
6	6.4684	6.633	6.8019	6.9753	7.1533	7.3359	7.5233	7.7156	7.9129	8.1152	8.3227	8.5355	8.7537
7	7.6625	7.8983	8.142	8.3938	8.654	8.9228	9.2004	9.4872	9.7833	10.089	10.405	10.73	11.067
8	8.8923	9.2142	9.5491	9.8975	10.26	10.637	11.028	11.436	11.859	12.3	12.757	13.233	13.727
9	10.159	10.583	11.027	11.491	11.978	12.488	13.021	13.579	14.164	14.776	15.416	16.085	16.786
10	11.464	12.006	12.578	13.181	13.816	14.487	15.193	15.937	16.722	17.549	18.42	19.337	20.304
11	12.808	13.486	14.207	14.972	15.784	16.645	17.56	18.531	19.561	20.655	21.814	23.045	24.349
12	14.192	15.026	15.917	16.87	17.888	18.977	20.141	21.384	22.713	24.133	25.65	27.271	29.002
13	15.618	16.627	17.713	18.882	20.141	21.495	22.953	24.523	26.212	28.029	29.985	32.089	34.352
14	17.086	18.292	19.599	21.015	22.55	24.215	26.019	27.975	30.095	32.393	34.883	37.581	40.505
15	18.599	20.024	21.579	23.276	25.129	27.152	29.361	31.772	34.405	37.28	40.417	43.842	47.58
16	20.157	21.825	23.657	25.673	27.888	30.324	33.003	35.95	39.19	42.753	46.672	50.98	55.717
17	21.762	23.698	25.84	28.213	30.84	33.75	36.974	40.545	44.501	48.884	53.739	59.118	65.075
18	23.414	25.645	28.132	30.906	33.999	37.45	41.301	45.599	50.396	55.75	61.725	68.394	75.836
19	25.117	27.671	30.539	33.76	37.379	41.446	46.018	51.159	56.939	63.44	70.749	78.969	88.212
20	26.87	29.778	33.066	36.786	40.995	45.762	51.16	57.275	64.203	72.052	80.947	91.025	102.44
21	28.676	31.969	35.719	39.993	44.865	50.423	56.765	64.002	72.265	81.699	92.47	104.77	118.81

22	30.537	34.248	38.505	43.392	49.006	55.457	62.873	71.403	81.214	92.503	105.49	120.44	137.63
23	32.453	36.618	41.43	46.996	53.436	60.893	69.532	79.543	91.148	104.6	120.2	138.3	159.28
24	34.426	39.083	44.502	50.816	58.177	66.765	76.79	88.497	102.17	118.16	136.83	158.66	184.17
25	36.459	41.646	47.727	54.865	63.249	73.106	84.701	98.347	114.41	133.33	155.62	181.87	212.79
26	38.553	44.312	51.113	59.156	68.676	79.954	93.324	109.18	128	150.33	176.85	208.33	245.71
27	40.71	47.084	54.669	63.706	74.484	87.351	102.72	121.1	143.08	169.37	200.84	238.5	283.57
28	42.931	49.968	58.403	68.528	80.698	95.339	112.97	134.21	159.82	190.7	227.95	272.89	327.1
29	45.219	52.966	62.323	73.64	87.347	103.97	124.14	148.63	178.4	214.58	258.58	312.09	377.17
30	47.575	56.085	66.439	79.058	94.461	113.28	136.31	164.49	199.02	241.33	293.2	356.79	434.75
31	50.003	59.328	70.761	84.802	102.07	123.35	149.58	181.94	221.91	271.29	332.32	407.74	500.96
32	52.503	62.701	75.299	90.89	110.22	134.21	164.04	201.14	247.32	304.85	376.52	465.82	577.1
33	55.078	66.21	80.064	97.343	118.93	145.95	179.8	222.25	275.53	342.43	426.46	532.04	664.67
34	57.73	69.858	85.067	104.18	128.26	158.63	196.98	245.48	306.84	384.52	482.9	607.52	765.37
35	60.462	73.652	90.32	111.43	138.24	172.32	215.71	271.02	341.59	431.66	546.68	693.57	881.17
36	63.276	77.598	95.836	119.12	148.91	187.1	236.12	299.13	380.16	484.46	618.75	791.67	1014.3
37	66.174	81.702	101.63	127.27	160.34	203.07	258.38	330.04	422.98	543.6	700.19	903.51	1167.5
38	69.159	85.97	107.71	135.9	172.56	220.32	282.63	364.04	470.51	609.83	792.21	1031	1343.6
39	72.234	90.409	114.1	145.06	185.64	238.94	309.07	401.45	523.27	684.01	896.2	1176.3	1546.2
40	75.401	95.026	120.8	154.76	199.64	259.06	337.88	442.59	581.83	767.09	1013.7	1342	1779.1
41	78.663	99.827	127.84	165.05	214.61	280.78	369.29	487.85	646.83	860.14	1146.5	1530.9	2047
42	82.023	104.82	135.23	175.95	230.63	304.24	403.53	537.64	718.98	964.36	1296.5	1746.2	2355
43	85.484	110.01	142.99	187.51	247.78	329.58	440.85	592.4	799.07	1081.1	1466.1	1991.7	2709.2
44	89.048	115.41	151.14	199.76	266.12	356.95	481.52	652.64	887.96	1211.8	1657.7	2271.5	3116.6
45	92.72	121.03	159.7	212.74	285.75	386.51	525.86	718.9	986.64	1358.2	1874.2	2590.6	3585.1

Table A5. Present value of an ordinary annuity

years	3%	4%	5%	6%	7%	8%	9%	10%	11%	12%	13%	14%	15%
1	0.9709	0.9615	0.9524	0.9434	0.9346	0.9259	0.9174	0.9091	0.9009	0.8929	0.885	0.8772	0.8696
2	1.9135	1.8861	1.8594	1.8334	1.808	1.7833	1.7591	1.7355	1.7125	1.6901	1.6681	1.6467	1.6257
3	2.8286	2.7751	2.7232	2.673	2.6243	2.5771	2.5313	2.4869	2.4437	2.4018	2.3612	2.3216	2.2832
4	3.7171	3.6299	3.546	3.4651	3.3872	3.3121	3.2397	3.1699	3.1024	3.0373	2.9745	2.9137	2.855
5	4.5797	4.4518	4.3295	4.2124	4.1002	3.9927	3.8897	3.7908	3.6959	3.6048	3.5172	3.4331	3.3522
6	5.4172	5.2421	5.0757	4.9173	4.7665	4.6229	4.4859	4.3553	4.2305	4.1114	3.9975	3.8887	3.7845
7	6.2303	6.0021	5.7864	5.5824	5.3893	5.2064	5.033	4.8684	4.7122	4.5638	4.4226	4.2883	4.1604
8	7.0197	6.7327	6.4632	6.2098	5.9713	5.7466	5.5348	5.3349	5.1461	4.9676	4.7988	4.6389	4.4873
9	7.7861	7.4353	7.1078	6.8017	6.5152	6.2469	5.9952	5.759	5.537	5.3282	5.1317	4.9464	4.7716
10	8.5302	8.1109	7.7217	7.3601	7.0236	6.7101	6.4177	6.1446	5.8892	5.6502	5.4262	5.2161	5.0188
11	9.2526	8.7605	8.3064	7.8869	7.4987	7.139	6.8052	6.4951	6.2065	5.9377	5.6869	5.4527	5.2337
12	9.954	9.3851	8.8633	8.3838	7.9427	7.5361	7.1607	6.8137	6.4924	6.1944	5.9176	5.6603	5.4206
13	10.635	9.9856	9.3936	8.8527	8.3577	7.9038	7.4869	7.1034	6.7499	6.4235	6.1218	5.8424	5.5831
14	11.296	10.563	9.8986	9.295	8.7455	8.2442	7.7862	7.3667	6.9819	6.6282	6.3025	6.0021	5.7245
15	11.938	11.118	10.38	9.7122	9.1079	8.5595	8.0607	7.6061	7.1909	6.8109	6.4624	6.1422	5.8474
16	12.561	11.652	10.838	10.106	9.4466	8.8514	8.3126	7.8237	7.3792	6.974	6.6039	6.2651	5.9542
17	13.166	12.166	11.274	10.477	9.7632	9.1216	8.5436	8.0216	7.5488	7.1196	6.7291	6.3729	6.0472
18	13.754	12.659	11.69	10.828	10.059	9.3719	8.7556	8.2014	7.7016	7.2497	6.8399	6.4674	6.128
19	14.324	13.134	12.085	11.158	10.336	9.6036	8.9501	8.3649	7.8393	7.3658	6.938	6.5504	6.1982
20	14.877	13.59	12.462	11.47	10.594	9.8181	9.1285	8.5136	7.9633	7.4694	7.0248	6.6231	6.2593
21	15.415	14.029	12.821	11.764	10.836	10.017	9.2922	8.6487	8.0751	7.562	7.1016	6.687	6.3125

22	15.937	14.451	13.163	12.042	11.061	10.201	9.4424	8.7715	8.1757	7.6446	7.1695	6.7429	6.3587
23	16.444	14.857	13.489	12.303	11.272	10.371	9.5802	8.8832	8.2664	7.7184	7.2297	6.7921	6.3988
24	16.936	15.247	13.799	12.55	11.469	10.529	9.7066	8.9847	8.3481	7.7843	7.2829	6.8351	6.4338
25	17.413	15.622	14.094	12.783	11.654	10.675	9.8226	9.077	8.4217	7.8431	7.33	6.8729	6.4641
26	17.877	15.983	14.375	13.003	11.826	10.81	9.929	9.1609	8.4881	7.8957	7.3717	6.9061	6.4906
27	18.327	16.33	14.643	13.211	11.987	10.935	10.027	9.2372	8.5478	7.9426	7.4086	6.9352	6.5135
28	18.764	16.663	14.898	13.406	12.137	11.051	10.116	9.3066	8.6016	7.9844	7.4412	6.9607	6.5335
29	19.188	16.984	15.141	13.591	12.278	11.158	10.198	9.3696	8.6501	8.0218	7.4701	6.983	6.5509
30	19.6	17.292	15.372	13.765	12.409	11.258	10.274	9.4269	8.6938	8.0552	7.4957	7.0027	6.566
31	20	17.588	15.593	13.929	12.532	11.35	10.343	9.479	8.7331	8.085	7.5183	7.0199	6.5791
32	20.389	17.874	15.803	14.084	12.647	11.435	10.406	9.5264	8.7686	8.1116	7.5383	7.035	6.5905
33	20.766	18.148	16.003	14.23	12.754	11.514	10.464	9.5694	8.8005	8.1354	7.556	7.0482	6.6005
34	21.132	18.411	16.193	14.368	12.854	11.587	10.518	9.6086	8.8293	8.1566	7.5717	7.0599	6.6091
35	21.487	18.665	16.374	14.498	12.948	11.655	10.567	9.6442	8.8552	8.1755	7.5856	7.07	6.6166
36	21.832	18.908	16.547	14.621	13.035	11.717	10.612	9.6765	8.8786	8.1924	7.5979	7.079	6.6231
37	22.167	19.143	16.711	14.737	13.117	11.775	10.653	9.7059	8.8996	8.2075	7.6087	7.0868	6.6288
38	22.492	19.368	16.868	14.846	13.193	11.829	10.691	9.7327	8.9186	8.221	7.6183	7.0937	6.6338
39	22.808	19.584	17.017	14.949	13.265	11.879	10.726	9.757	8.9357	8.233	7.6268	7.0997	6.638
40	23.115	19.793	17.159	15.046	13.332	11.925	10.757	9.7791	8.9511	8.2438	7.6344	7.105	6.6418
41	23.412	19.993	17.294	15.138	13.394	11.967	10.787	9.7991	8.9649	8.2534	7.641	7.1097	6.645
42	23.701	20.186	17.423	15.225	13.452	12.007	10.813	9.8174	8.9774	8.2619	7.6469	7.1138	6.6478
43	23.982	20.371	17.546	15.306	13.507	12.043	10.838	9.834	8.9886	8.2696	7.6522	7.1173	6.6503
44	24.254	20.549	17.663	15.383	13.558	12.077	10.861	9.8491	8.9988	8.2764	7.6568	7.1205	6.6524
45	24.519	20.72	17.774	15.456	13.606	12.108	10.881	9.8628	9.0079	8.2825	7.6609	7.1232	6.6543

Glossary

Accrual system of accounting: An accounting method in which revenues are recognized when earned rather than received, and expenses are recognized when incurred or charged rather than when paid.

Agribusiness: The activities of supplying goods and services to growers and ranchers providing food and fiber products, and the marketing of these products to the end customers, households and businesses.

Agricultural mortgage-backed security: A bond-type security that is backed by a pool of mortgages as collateral, with the further guarantee of Farmer Mac. Usually labeled AMBS.

Amortization: The mechanics of repayment of the loan, often in equal payments over the loan term.

Amortization schedule: A table that details the payments, balance, interest paid, and reduction in principal for a promissory note.

Amortizing, fully: Periodic loan payments are sufficient to extinguish the debt (pay off the entire principal) over the term of the loan.

Amortizing, partially: Periodic loan payments make some reduction in the principal balance but do not fully extinguish the debt (do not pay off the entire principal) over the term of the loan.

Annual percentage rate: The true interest rate for the loan, found by using time value of money calculations to determine the actual yield to the lender. Usually referred to as the APR.

Annuity: A series of equal, periodic cash flows over a finite life.

Annuity due: An annuity in which the cash flows occur at the beginning of each period.

Assets: Economic resources owned by a business, either tangible or intangible.

Balance sheet: A financial statement that reports the stock value of the assets, liabilities, and owner's equity of a business on a specific date, usually the end of a fiscal period.

Balloon payment: A lump-sum payment required at the due date of the promissory note, usually the remaining unpaid principal (balance).

Beneficiary: The lender; the party that will be repaid in a promissory note.

Benefit-cost ratio: The ratio of the sum of the present value of cash inflows divided by the sum of the present value of the cash outflows.

Business risk: The uncertainty or variation in income or returns of a business over time due to the nature of its business.

Capital asset: An asset with an economic life of more than one year.

Capital budgeting: The process of planning expenditures on assets whose returns will extend beyond one year.

Capital lease: Long-term lease that is noncancelable, in which rent is paid at least annually.

Capital markets: The set of institutions and participants that makes a market for long-term securities.

Capital replacement and term debt repayment margin: Net farm income from operations plus miscellaneous revenue minus miscellaneous expense plus nonfarm income plus depreciation expense minus income tax expense minus owner withdrawals minus payment on unpaid operating debt from a prior period minus principal payments on the current portion of term debt minus principal payments on the current portion of capital leases minus total annual payments on personal liabilities. A measure of repayment capacity.

Cash budget: An informal financial statement prepared by the business to forecast future cash flows and the need for a cash line of credit from a bank.

Cash lease: The rent in this lease is paid in cash, usually in advance.

Cash system of accounting: An accounting method in which revenues are recognized when received and expenses are recognized when paid.

Cash window: A means of selling loans to Farmer Mac that allows delivery of the loan documents from within seven days to eight weeks.

Casualty risk: The risk to a business of an earthquake, accident, injury, fire, or other physical disaster.

Chicago Board of Trade: The first commodities exchange, established in the nineteenth century, to make a market for agricultural products.

Commercial bank: Depository financial intermediary that invests mainly in commerical loans, including agribusiness loans.

Common-size statement: Financial statements expressing each account balance as a percentage of a designated total.

Compensating factors: Financial and personal issues that positively affect the likelihood that the borrower will meet the borrower's obligation, to justify a favorable ruling by the underwriter on the loan application.

Compound interest: Interest added to the loan principal, which from that point forward earns interest too.

Contract for deed: Also known as the land contract; a form of installment financing for real estate, similar to automobile financing, in which the title is transferred to the buyer upon payoff of the entire promissory note.

Cost of capital: Cost of financing a business expressed as an interest rate; the weighted average cost of a business's debt and equity capital.

Cost of goods sold: The value of inventory (including current production) that has been sold by the agribusiness for a specific time period, usually labeled COGS.

Credit reserves: Additional borrowing capacity, available for "rainy day" borrowing.

Credit terms: The contractual agreements between trade credit parties, describing the timing of payments and any discounts.

Current assets: Short-term assets that can be utilized within one year.

Current liabilities: Debts due within one year.

Current ratio: Current assets divided by current liabilities. A measure of liquidity.

Deed of trust: A transient or temporary deed granting title to a third party, the trustee, if the borrower fails to meet the terms of the promissory note and goes into default.

Default: Failure of the borrower to meet the terms of the promissory note; usually failure to make timely payment.

Discounting: The time value of money process of finding the value today of some future cash flow(s), given a discount rate.

Discount rate: (1) The formal interest rate at which the Federal Reserve Bank lends to borrower banks. (2) The market interest rate for a debt security such as a bond or a mortgage promissory note. (3) Any interest rate for a specific asset-pricing problem.

Diversification: Distributing the firm's assets over several crops, products, or enterprises to reduce business risk.

Dominating (or efficient) set: The investments or portfolios of investments that maximize the returns to the investor for all possible levels of business risk.

Dual chartering: The unique system of bank and savings and loan licensing found in the United States, whereby those institutions can obtain either state or federal charters to operate.

Economic risk: The risk that macroeconomic variables can impact the business.

Equity capital: The value of the owner's investment, which usually includes current- and prior-period earnings that have not been withdrawn from the business.

Escrow: The process in which a third party to a transaction holds documents and funds until specified conditions have been met.

Excess paid-in capital: The owner's equity account that shows the dollars above the par value of stock that were raised by a corporation when the stock was sold.

Expense: Cost incurred doing business.

Farm cooperative: A not-for-profit corporation that provides a service to its members by collectively representing them in the sale of their production.

Farm Credit System: Legislatively created farm credit agencies that meet a portion of the borrowing needs of domestic agriculture.

Farmer Mac: Federal Agricultural Mortgage Corporation, a Government Sponsored Enterprise that provides secondary-market financing for farm, ranch, and rural home mortgages.

Farm Financial Standards Council: A volunteer group of ag producers, accountants, lenders, trade organization representatives, and academicians organized in 1989 to clarify and standardize agricultural financial statements.

Farm Service Agency: The 1994 replacement for the Farmers' Home Administration, a farm and rural home-loan lender.

Fed Funds rate: A short-term lending rate between banks for use in meeting reserve requirements. One bank lends its excess reserves to another bank that is temporarily short of reserves.

Finance: The study of the flow of funds in an economy or firm.

Financial efficiency: A manager's ability to control costs and utilize assets efficiently.

Financial risk: The risk of losing equity capital and credit reserves during periods of adverse conditions.

Financing statement: A summary of property that is taken as collateral. It is recorded to provide public notice of a lien against the property.

Foreclosure: The legal process of recovering the real estate collateral when the borrower is in default of the promissory note.

Forward contracting: Selling or buying a commodity or security with the price set at the beginning of the contract, but with delivery of the asset to take place at the contract maturity.

Futures contract: An agreement with the price set in advance to make or take delivery of a quantity of a commodity at or before the contract maturity, but where the parties can forgo delivery by reversing their trades at the end.

Futures options contracts: Options contracts on futures contracts.

Future value: The value in the future of one or a series of cash flows, invested at some interest rate.

GAAP: Generally accepted accounting principles, as prescribed by the Financial Accounting Standards Board (FASB) and given the blessing of the Farm Financial Standards Council (FFSC).

Gain (or loss) on the sale of a capital asset: The difference between the sales price of a capital asset and its book value (cost minus accumulated depreciation).

Going long: The purchase of a futures contract at the beginning.

Government Sponsored Enterprise: A business, frequently owned by public stockholders, that was originally established by Congress for the public good. Known as GSEs, examples include Farmer Mac and Fannie Mae (Federal National Mortgage Association).

Hedging: Taking a position in a security or asset equal to but opposite to one's existing position in another asset, to minimize price risk.

Historical cost principle: The accounting principle that says that assets appear on the balance sheet at their original cost, or cost minus accumulated depreciation.

Holding company: A separate corporation that owns other corporations for the purposes of providing financial "firewalls" between them, most often found in commercial banking.

Horizontal analysis: Financial statement analysis that looks closely at how values of each account change from period to period, also known as trend analysis.

Human risk: The risk that human behavior, including employee health, will adversely affect the business.

Income statement: The financial statement that reports current period operations, measuring revenues, expenses, and net income (loss).

Individual retirement account: Congressionally created private retirement plans that allow the participant to make annual investments that are generally tax deductible, usually labeled an IRA.

Interest: The money to which the lender is entitled as revenues from a loan, based on the terms of the promissory note.

Interest rate lock: The contractual agreement to make or sell a loan at a specific interest rate.

Intermediary: Financial institution that accepts deposits or investments, makes other investments, and profits on the spread of the interest rates.

Internal rate of return: The discount rate at which the sum of the present value of the cash inflows equals the sum of the present value of the cash outflows; the compound interest rate earned by an investment.

Intraperiod compounding: Interest is calculated and added to the principal more frequently than once per year. For loans, monthly compounding is most common (twelve times per year).

Investment banking: The activities of stock brokerage firms, including the initial underwriting of securities.

Land contract: A method of financing real estate where the title passes to the buyer only after all payments have been made. Also called the contract for deed.

Lease: Contractual agreement between lessor (landlord) and lessee (tenant) for the use of real or personal property, with the lessee paying rent to the lessor.

Legal risk: The risk that the business will face lawsuit.

Lender forbearance: Goodwill efforts on the part of the lender to work with a distressed borrower.

Leverage: The degree to which a business is financed by debt instead of owner's equity.

Liabilities: Debts owed by the business.

Liquidity: The ability to meet short-term obligations with short-term assets.

Loan servicing: The maintenance of payment, tax, and other records for the lender and borrower of a loan.

Loan term: The length of time the borrower has to repay the loan.

Loan-to-value ratio: The ratio of the amount borrowed or current loan balance to the appraised value of the asset, usually referring to real-estate borrowing.

MACRS: The modified accelerated cost recovery system, a depreciation method enacted by Congress for tax depreciation purposes.

Maintain savings: A risk management technique in which the farmer keeps cash in savings accounts to cover any negative cash flows during bad times.

Maker of a note: The individual(s) who is borrowing and signs the promissory note as an agreement to repay the indebtedness.

Margin: The funds left over after the business has made all payments on debt and capital leases.

Matching principle: The accounting principle that says that revenues should be paired up on an income statement with expenses that generated those revenues.

Middleman: The financial institution that earns revenue by the commissions received from financial transactions.

Monetary policy: The actions taken by the U.S. Federal Reserve Bank to manage the nation's money supply.

Mortgage: A pledge of real estate to a lender as security (collateral) for a debt.

Mortgagee: The lender on a mortgage in mortgage states.

Mortgagor: The borrower on a mortgage in mortgage states.

Mutual fund: Financial intermediary that acts as an investment company, selling shares of stock to investors, then investing the proceeds on their behalf in other investments such as stocks or bonds.

Net farm income from operations: Gross revenues minus operating and interest expenses, usually labeled NFIFO.

Net present value method: A sophisticated capital budgeting method that determines whether the asset generates sufficient cash flows, discounted at the cost of capital, to pay for itself during its useful life. Usually labeled the NPV.

Net yield: The interest rate at which Farmer Mac discounts mortgages at its cash window, also labeled the required net yield.

Noncurrent assets: Assets having economic lives of greater than one year.

Noncurrent liabilities: Debts to be paid in more than one year's time.

Obsolescence risk: Using or adopting technologies than become outdated before they provide positive returns.

Operating lease: Short-term lease in which total rent is based upon how much time the lessee uses the property.

Operating-profit-margin ratio: The quantity (net farm income from operations plus farm interest expense minus owner withdrawals for unpaid labor and management) divided by gross revenues. A measure of profitability.

Option contract: The right but not the obligation to buy (call) or sell (put) a security or other asset at a specified exercise (strike) price during the option period.

Ordinary annuity: Annuity in which the cash flow occurs at the end of each period.

Owner's equity: The net worth of a business.

Partial amortization: Repayment of a loan that does not fully pay off over the required loan term, therefore requiring a balloon payment or other action when it is due.

Payback method: A simple capital budgeting method where the budgeter determines the number of years in which an investment will recoup its initial cost by generating net cash inflows.

Points: Loan fees that are viewed as prepaid interest and raise the APR of the loan. One point is 1% of the loan amount.

Political risk: Actions by domestic or foreign government bodies that can impact the business.

Portfolio: A set of investments.

Preference decision: Comparing and ranking several capital budgeting projects.

Present value: The discounted value today of one or a series of cash flows.

Price risk: Uncertainty in returns due to variations in the market prices of inputs and outputs.

Principal: Usually, the balance of the loan; the amount owed.

Production risk: Uncertainty in returns due to variations in quantity of output.

Profitability: The measurement of profits in relation to assets, equity, or gross revenues.

Promissory note: Written evidence and description of a debt, including the names of the borrower and lender, the borrower's promise to pay, the interest rate, the term, the frequency of payments, and the naming of collateral.

Rate of return on farm equity: The quantity (net farm income from operations minus owner withdrawals for unpaid labor and management) divided by the average total farm equity. A measure of profitability.

Refinancing: The paying off of an existing loan with funds received from a new loan, usually undertaken to lower the interest rate paid.

Regulation Z: A by-product of the 1968 Truth in Lending Act, which required the Federal Reserve Bank to devise a means of reporting APRs to borrowers. The Fed wrote Regulation Z to meet this requirement.

Repayment capacity: The ability of a borrower to pay principal and interest on loans as well as meet all other cash obligations.

Required reserves: The dollar amount of cash and deposits at the Federal Reserve Bank that a commercial bank or savings and loan must hold as an asset to meet Fed guidelines.

Retained surplus: The term used in bank financial statements that means current and prior period net income not paid out as dividends. It is synonymous with retained earnings or earned surplus.

Revenue: Money earned by a business.

Reversing trade: The trade executed by a futures contract party at the end of his or her holding period; it is the opposite trade to that executed at the beginning.

Risk aversion: Investor's preference to avoid risk.

Risk-free investment: An investment that has no business risk; returns are stable.

Screening decision: Determining whether a capital budgeting project meets or passes an initial standard such as a positive NPV or IRR greater than the cost of capital.

Secondary market: The set of institutions that makes a market for the resale of securities, including the promissory note within a mortgage.

Securitization: The process of issuing securities backed by pools of loans;

the principal and interest paid on the loans is used to pay the principal and interest due on the securities.

Security: Written or electronic evidence of a claim on the assets or income of the issuer.

Security agreement: A standardized description of assets pledged as collateral for a debt.

Selling short: The sale of a futures contract at the beginning of the trade.

Share lease: Rent is paid upon harvesting the crop with the landlord receiving some predetermined portion of the output.

Simple rate of return: The total net income provided by an asset divided by its initial cost.

Solvency: The degree to which all assets exceed all liabilities.

Spreading sales: Selling production throughout the production season rather than all at once, to minimize price risk.

Statement of cash flows: A financial statement that summarizes the changes to the cash account over the accounting period.

Statement of owner's equity: The financial statement that shows changes between beginning and ending owner's equity for an accounting period.

Stay of foreclosure: A delaying tactic by a borrower who files bankruptcy, which forces the court handling the suit of foreclosure to postpone the suit until the bankruptcy is resolved.

Stock value: The value of an account at a point in time.

Subordination clause: A statement in a promissory note or mortgage that requires the lender to allow another lender to take a priority claim on the mortgage collateral, often used in real-estate development.

Suit of foreclosure: The formal lawsuit by the lender to obtain the property.

Swap: The exchange of a pool of loans for Farmer Mac's AMBSs.

Term debt and capital lease coverage ratio: The quantity (net farm income from operations plus miscellaneous revenues minus miscellaneous expense plus nonfarm income plus depreciation expense plus interest on term debt plus interest on capital leases minus income tax expense minus owner withdrawals) divided by the quantity (annual scheduled principal and interest payments on term debt plus annual scheduled principal and interest payments on capital leases). A measure of repayment capacity.

Time value of money: The universal preference for a dollar today versus a dollar at some future point in time.

Trade credit: Short-term financing provided by a vendor.

Trustee: The independent third party that enforces the deed of trust should the borrower go into default on the promissory note.

Trustor: In deed-of-trust states, the borrower in a real-estate loan.

Underwriter: The expert individual or software that determines whether the borrower is capable of repayment of the requested loan and makes a decision on the application.

Vendor: Merchant or business selling products to businesses and the public.

Working capital: The difference between current assets and current liabilities.

Index